D0554147

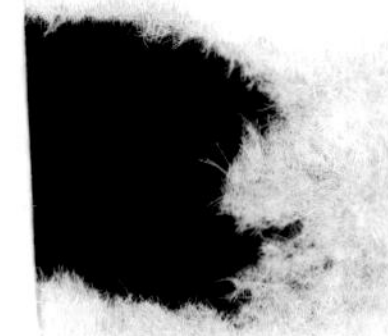

WORLD ENERGY RESOURCES 1985–2020

WORLD ENERGY RESOURCES 1985–2020

Executive Summaries of reports on resources, conservation and demand to the Conservation Commission of the World Energy Conference

Published for the WEC by
IPC Science and Technology Press

ISBN 0 902852 90 6 (hard cover)
ISBN 0 902852 91 4 (paperback)

Published simultaneously in the UK and the USA by IPC Science and Technology Press, IPC House, 32 High Street, Guildford, UK, GU1 3EW, and 205 East 42nd Street, New York, NY 10017, USA

Printed in England by Kingprint Ltd, Orchard Road, Richmond, Surrey.

CONTENTS

vii Foreword

xi Conservation Commission

xii Review Boards

1 Worldwide petroleum supply limits

49 The future for world natural gas supply

57 An appraisal of world coal resources and their future availability

87 Hydraulic resources

105 The contribution of nuclear power to world energy supply

135 Unconventional energy resources

181 Energy conservation

213 World energy demand

249 National Committees of the World Energy Conference

The International Executive Council of the World Energy Conference wishes to express its appreciation of the inspiration, tireless energy and leadership of the Chairman of the Conservation Commission which has led to the creation of these Reports; also the gratitude of its members to the John and Margaret Lee Kiely Foundation for its generous financial support of the work of the Commission.

The Commission was set up and the major part of the study was completed during the time when my predecessor, Monsieur Roger Gaspard, was Chairman of the International Executive Council. I have closely followed its progress and I wish to express our debt of gratitude to the authors of the Reports, to the members of the Commission and to our Secretary-General for the time and effort which they have put into the realization of this Survey.

I should also like to recognize the role of the Chairmen and members of the Review Boards and Panels for their constructive contribution before and during the Tenth World Energy Conference in Istanbul in providing creative criticism of these reports.

Heinrich Mandel
Chairman of the International Executive Council
World Energy Conference

FOREWORD

The material contained in this publication consists of Executive Summaries of reports made to the Conservation Commission of the World Energy Conference. The Summaries were first released at the Tenth World Energy Conference in Istanbul, in September 1977, as documents which formed the basis of Round-table discussions of energy demand, supply, and conservation. They are a unique set of papers, since they include the first collection of views of the world's main practitioners in the energy field on the prospects for world energy supply during the next four decades.

The Summaries, together with the Study Group Reports in full, now constitute the main working papers for the preparation of a report by the Conservation Commission to the World Energy Conference. Recent work by other energy groups is also being carefully weighed by the Commission in the preparation of its report. When issued, the report will present the Commission's views on energy demand, resources, conservation, research and development, and, finally, energy strategies.

Early in its deliberations, the Commission decided it would not produce its report until the contents of the Study Group Reports had been exposed to critical review by those throughout the world who are knowledgeable in the energy field. The review process has included critiques by representatives of inter-governmental organizations, discussions with member National Committees of the World Energy Conference, examination of the Executive Summaries by special Review Boards, and, finally, the Round-table discussions at the Istanbul Conference in September 1977.

The World Energy Conference is a 54-year-old organization comprising 76 National Committees representing as many nations. The purposes of the World Energy Conference are to promote the development and the peaceful use of energy resources to the greatest benefit of all, both nationally and internationally, by:

(i) considering the potential resources and all the means of production, transportation, transformation, and utilization of energy in all their aspects;
(ii) considering energy consumption in its overall relationship to the growth of economic activity in the area;
(iii) collecting and publishing data on the matters mentioned in (i) and (ii);
(iv) holding conferences of those concerned in any way with the matters mentioned in (i) and (ii).

These purposes are implemented through the interaction of National Committees which are broadly representative of the energy planning, supply, conversion, utilization, and conservation interests of each nation.

The need for the Conservation Commission was recognized as early as 1972, when it was proposed that the World Energy Conference consider the matters of world energy resources, demand, and conservation in the light of what appeared to be the development of potential shortfalls in energy availability. In its terms of reference, when formally established in 1975, the Commission was asked to consider, for the period 1985 to 2020, ways in which the supply of energy from non-renewable and renewable energy resources might be improved, as well as to assess the opportunity for the conservation of oil and gas by the substitution of other, more plentiful, renewable and non-renewable resources and the probable extent to which conservation measures might reduce total energy demand.

To obtain the necessary factual information on energy resources and techniques of utilization from which it could develop its view of the world energy prospect, the Commission asked certain of the National Committees of the World Energy Conference to provide authoritative reports on the supply of oil, gas, coal, and nuclear energy; unconventional energy resources; and conservation measures. Those that were selected to provide reports on energy supply formed Study Groups including internationally prominent organizations. In addition, the Conservation

Commission engaged the services of the Energy Research Group of the Cavendish Laboratory of Cambridge University to produce a report on projected energy demands for the period under review.

The Commission met regularly to review the progress of the Study Groups and to offer advice and guidance. It was, of course, essential that the reports reflect as much as possible expert international opinion. To achieve this, the various Study Groups adopted different techniques, each of which was considered most appropriate for the particular energy resource. For example, the Study Group for Oil polled international experts using the Delphi method; the Study Group for Coal obtained information from national authorities by questionnaire; and the Study Group responsible for Uranium Resources reviewed its work with the members of the NEA–IAEA Steering Group for Uranium Resources. In addition, members of the Study Groups consulted energy authorities in various parts of the world to discuss the assembled data and conclusions reached by their groups.

As stated earlier, the Commission decided that the Tenth World Energy Conference to be held in Istanbul in September 1977, would provide an exceptional opportunity to obtain an even broader sounding of expert opinion from around the world. The Turkish National Committee, host to the Conference, was very receptive to this idea. Accordingly, in conjunction with their Organizing Committee, it was arranged to hold four Round-table discussions at the Conference; one on Demand and Conservation and the others on Oil and Gas, Coal, and Nuclear and Unconventional (solar, geothermal, etc.) energy resources.

In preparation for effective Round-table discussions, Review Boards of world-renowned experts in the field of resources, supply, demand, and conservation were convened early in July 1977 to comment on the material available at that time and to assist the chairmen of the Round-tables in the development of a Round-table discussion format. As a result of the comments received at these sessions, the authors made some modifications to their reports, and the Review Boards, further, identified issues for discussion at the forthcoming Istanbul Round-tables.

There was no thought, prior to the Conference, that the reports produced by the Study Groups would be issued in advance of the report to be produced by the Commission itself. At the Conference, however, there were many representations calling for an early publication of these documents for three reasons: the quality of the work, the authoritativeness of the sources, and the urgency for action recognized by many and brought out in several of the Summaries themselves. The Commission has, accordingly, in response to these representations, decided to release as separate publications the Executive Summaries and subsequently the reports.

As a matter of deliberate policy, these reports were prepared quite independently. They were produced as working material for the Conservation Commission and still require reconciliations and interpretation.

(a) The Resource Studies are the contributions of expert practitioners and represent authoritative opinions on the prospects of energy supply in each particular field. However, the Commission will examine, in the light of additional information, the various possible production rates for primary energy resources up to the year 2020. For example, unconventional petroleum resources (heavy oils, oil shales, tar sands, and deep offshore deposits) bear further study, since they may well constitute a stabilizing element in the world oil market by the year 2000. In the case of coal, if firm supply contracts and attractive pricing were to develop, it is conceivable that production rates by 2020 could significantly exceed the production rate projected in the Coal Study. There are factors related to other non-renewable and renewable resources, as well, which the Commission will consider prior to the publication of its position on primary energy resources.

(b) The conservation measures set out in the Conservation Executive Summary indicate that the potential for reductions in energy demand, through more efficient practices, is significant. However, there is some question about the proportion of these which can be properly deducted from an extrapolation of current trends, because these trends already include increasing efficiency in energy utilization. In addition, there is the question of whether all the energy savings apparently available through the proposed structural changes and improvements in energy utilization efficiencies can be achieved without stringent regulations. Conservation is a very complex matter, and the Commission will consider further the feasibility of the measures proposed by the Study Group before arriving at an opinion on the possible effect of energy conservation measures to the year 2020.

(c) An assessment of future possible energy supply has little meaning without a corresponding projection of likely demand. It was for this reason that the Conservation Commission engaged the Energy Research Group of Cambridge to produce a report on energy demand for the period 1985 to 2020. This is the Commission's basic reference on prospective demand; however, other sources of relevant demand information will be consulted. Energy demand, reciprocally involved as it is with demographic, social, economic, and political development, is impossible to predict accurately over even relatively short periods, let alone a period extending nearly half a century into the future. In treating this complicated matter, the Energy Research Group found it necessary to develop a method of estimating forward demand. In its work, the Commission will want to weigh some of the other possibilities before arriving at its conclusions. Also, the Group, in order to produce reasonable demand projections by energy sector in selected geo-economic regions of the world, developed several scenarios based on various assumptions regarding economic growth, energy pricing, and conservation. However, the authors emphasized that these scenarios are not forecasts; nor do they represent the Conservation Commission's view of the future. Nevertheless, they provided the basis for credible demand projections and are valuable in conveying an appreciation of the complexity of the matter. Scenarios are included with 'constrained demand', where constraint goes much beyond what is normally intended by the term 'conservation' and implies severe rationing and the adverse social and economic concomitants. It must be emphasized that, although the possibility of such an outcome cannot be dismissed, there is no acceptance by the Commission of its inevitability.

(d) There is an inherent difficulty in treating energy in various forms collectively. The various forms cannot be brought to a common base for summation or comparison for practical purposes in any simple way. In the fields of petroleum and uranium resources, for example, the total energy content of the petroleum resources and uranium resources may be expressed in terms of a common energy unit. This would not be very helpful, however, in relating the energy available for use from these two primary resources, since the fraction of the total energy available through current, or even prospective, technology is very different for each. Any attempt to treat them collectively implies assumptions, on the one hand, about the average heating value of the oils comprising the petroleum resource and, on the other hand, about the process for conversions of the nuclear binding energy of the uranium to other energy forms. Similarly in the field of secondary energy (energy delivered to the consumer), the energy in electricity has a very different value from that in fossil fuels. Electricity, for example, besides being essential for many important specific uses, is converted to heat or work in many applications at efficiencies that could not be approached by fossil fuels. Therefore, even though the Demand Study Report presents various forms of primary and secondary energy in single tables and graphs, the limitations mentioned here and the authors' qualifying remarks must be borne in mind.

This, then, is the origin and nature of the Executive Summaries. The Commission's consideration of the balance between resources and demand, based on these reports, is proceeding, and its findings will be presented later in the year. However, what is abundantly clear, from even a superficial reading of the Executive Summaries, is that positive decisions and appropriate actions are required now to avoid the risk of serious energy deficiencies with their grave social consequences in the last decade of this century and beyond. There are large coal reserves and uranium resources, to be sure; certain of the renewable resources (solar and geothermal) may make an increasingly important contribution to energy supply; and fusion may enter the commercial feasibility stage by 2020. However, the development of any of these on the scale necessary and in the time available requires immediate, positive action. This need is emphasized repeatedly in the Study Group Reports. The authors of the Report on Coal Resources state:

> The world possesses abundant coal occurrences. One may also assume, in addition to currently estimated resources and reserves, that there is a considerable 'potential behind the potential' . . . but . . . Considering the long lead-times required for an extensive production and use of coal (and this includes policy decisions, investment commitments, coal mine development, transportation, etc.), one cannot merely rely on a future market which might be more favourable to coal;

the authors of the Report on Uranium Resources conclude:

> Although nuclear energy is expected to provide an increasing share of the world's total energy supply over the study period, its share of the electrical energy component will not likely exceed an estimated 50 to 60%. It is clear that providing even this fraction of future world energy supply presents a challenge calling for unprecedented levels of international cooperation. In this context, urgent action is required to promote the development of both uranium resources and nuclear technology;

the authors of the Unconventional Energy Resources Report point out:

> The important point is that even if this [*a rate of fractional penetration of the energy market higher than that which occurred historically for natural gas*] could happen, it would require the better part of a century for a mix of solar-energy systems (assuming technical success, economic competitiveness with other energy options, and total social acceptability) to displace 50% of the United States' energy demand;*

and the main conclusion of the Report on Demand is:

> . . . high world economic growth during the period to 2020 would require early and successful action towards energy conservation and constraint accompanied by vigorous development of all forms of energy supply.

The objective of the forthcoming Conservation Commission Report will be to review prospects for energy supply and demand and to indicate general strategies to ensure the energy necessary to sustain continued improvement in the welfare of mankind. The Report cannot propose one overall energy policy for all nations; nor will it attempt to define the energy needs and demands of individual nations. However, it is hoped that each nation will find the Report to be of substantial value in preparing its own energy policy in the light of its particular economic, technological, and environmental requirements. Since energy is not only of national concern but also international, it will be necessary for the nations of the world to cooperate to an even greater extent with respect to the use of energy, trade in energy resources, and exchange of technology.

The World Energy Conference is contemplating continuation of the work of the Conservation Commission beyond the issue of its Report. Advice and comments from those with an interest in any aspect of the world's use of energy would be welcomed by the World Energy Conference.

FINAL DRAFT
29 December 1977

* Although the author of the Unconventional Energy Resources report is making this point with particular reference to the United States, it would be generally relevant to any industrial region.

CONSERVATION COMMISSION

Chairman
Mr John R. Kiely, Bechtel Corporation, USA

Members—National Committees
Dr G. Obermair, Ministry for Trade, Commerce & Industry, Austria
Monsieur L. de Heem, Association Vinçotte, Belgium
Monsieur P. Ailleret, Union Technique de l'Electricité, France
Dr J. S. Foster, Canada (Vice-Chairman of Commission)
Professor T. R. Gerholm, University of Stockholm, Sweden
Professor L. Heller, Institute of Energy Economy, Hungary
Mr W. A. L. Thomas, Federal Ministry of Mines & Power, Nigeria
Dr Călin Mihăileanu, Central Energy Research Institute, Romania
Mr Goro Inouye, Atomic Energy Commission, Japan
Shri A. K. Ghose, State Planning Board, India (and **Miss Otima Bordia**)

Members—International Organizations
Mr H. B. Chenery, International Bank for Reconstruction & Development, USA (and **Mrs Helen Hughes**)
Professor I. D. Stancescu, Centre for Natural Resources, Energy & Transport, United Nations
Mr E. F. Janssens, Economic Commission for Europe, Switzerland

Ex-Officio Members
Chairman of International Executive Council, **Dr H. Mandel,** Federal Republic of Germany
Chairman of the Programme Committee, **Dr S. Hultin,** Finland

Consultant
Dr R. J. Eden, Cambridge University Energy Research Group, UK

Policy Adviser
The Lord Hinton of Bankside, Electricity Supply Research Council, UK

Secretary-General
Mr E. Ruttley, World Energy Conference, UK

Resources Group
Co-ordinator—Professor T. R. Gerholm (Sweden)*

OIL
Dr P. Desprairies, Institut Français du Pétrole, France
Monsieur J. le Duigou, Institut Français du Pétrole, France

GAS
Mr L. J. Clark, formerly with International Gas Union, UK
Mr L Fish, American Gas Association, USA

COAL
Professor Dr H. Michaelis, Belgium
Dr G. Ott, German Coal Association, Federal Republic of Germany
Professor Dr W. Peters, Bergbau-Forschung GmbH, Federal Republic of Germany
Dr H. D. Schilling, Bergbau-Forschung GmbH, Federal Republic of Germany

HYDRAULIC
Dr Ellis L. Armstrong, USA

NUCLEAR ENERGY
Mr J. S. Foster*
Dr M. F. Duret, Atomic Energy of Canada Ltd, Canada
Mr R. M. Williams, Department of Energy, Mines & Resources, Canada

UNCONVENTIONAL RESOURCES
Professor P. L. Auer, Cornell University, USA

CONSERVATION STUDY
Mr L. Nevanlinna, Imatran Voima Osakeyhtioe, Finland
Mr F. Kommonen, EKONO, Finland

* Members of the Conservation Commission and the Commission's Resources Group.

REVIEW BOARDS

OIL AND GAS REVIEW BOARD

Chairman: *Mr W. A. Roberts, Vice-President, Phillips Petroleum Co. (USA)

Vice-Chairman: *Prof. Dr I. Kafesçioğlu, Vice-President, Turkish Petroleum Corporation (TURKEY)

Members:
*Mr A. Attiga, Secretary-General, OAPEC (KUWAIT)
*Mr M. M. Pennell, CBE, Deputy Chairman, British Petroleum Co. Ltd (UK)
*Mr A. R. Martinez, Permanent Commission, National Energy Council (VENEZUELA)
*Mr J. W. Kerr, Chairman, Trans Canada Pipelines (CANADA)
Dr J. Heubler, Senior Vice-President, Institute of Gas Technology (USA)
*Mr W. N. Sande, Syncrude Canada Ltd (CANADA)
Mr L. J. Clark, Immediate Past President, International Gas Union
Mr E. Holtermann, Secretary-General, Norsk Petroleumsinstitut (NORWAY)
Mr M. Lovegrove, British National Oil Corporation (UK)

Authors:
Dr P. Desprairies, Institut du Pétrole (FRANCE)
Mr W. McCormick, Jr, Vice-President, American Gas Association (USA)

COAL REVIEW BOARD

Chairman: *Dr-Ing. Dr rer. Pol. K. Bund, Vorsitzender des Vorstandes der Ruhrkohle AG (FRG)

Vice-Chairman: *Ass. Prof. Dr M. Ayan (TURKEY)
*Mr E. R. Phelps, President, Peabody Coal Co. (USA)

Members:
*Sir Derek Ezra, MBE, Chairman, National Coal Board (UK)
*Mr M. W. Sweetland, Director, Minerals Division, The BHP Co. Ltd (AUSTRALIA)
*Shri S. K. Bose, Joint Secretary, Department of Coal, Ministry of Energy (INDIA)
Dr D. J. Kotze, Director Energy, Department of Planning & Environment (SOUTH AFRICA)
*Mr J. H. Harvie, Vice-President, Kaiser Resources Ltd (CANADA)
*Dr U. Lantzke, Executive Director, International Energy Agency
*Dr-Ing. C. Lares Cordero, President, CA Minas de Naricual (VENEZUELA)
Dr G. B. Fettweis, Vorstand des Instituts für Bergaukunde, Montanuniversität Leoben (AUSTRIA)
Prof. Dr G. Bischoff, Universität Köln (FRG)

Authors:
Dr W. Peters (FRG)
Dr H. D. Schilling (FRG)

Authors of Reports as shown attended meetings of the Review Boards.

* Members of the Review Boards who served on the respective panels for the Round-table sessions at the Tenth World Energy Conference, Istanbul, September 1977 are shown with an asterisk.

NUCLEAR AND UNCONVENTIONAL ENERGY RESOURCES REVIEW BOARD

Chairman: *Sir John Hill, Chairman, United Kingdom Atomic Energy Authority

Vice-Chairman: *Prof. Dr N. Vejiroğlu (TURKEY)

Members:
*Mr E. Grafström, Chairman, Swedish State Power Board (SWEDEN)
*Sir John -Symmonds, Australian Atomic Energy Commission
*Mr J. A. Wilberg, Eletrobras (BRAZIL)
*Shri J. C. Shah, Chairman, Atomic Power Authority (INDIA)
*Mr D. G. S. Chuah, LAAS (CNRS) (MALAYSIA)
*Prof. Ing. T. Leardini, Direttore Centrale, Ente Nazionale Energia Elettrica (ITALY)
*Dr J. A. Phillips, Assistant Director General, International Atomic Energy Agency
*Dr M. Davis, Nuclear Energy, Electricity & New Energy Sources, Commission of the European Community
*Mr B. T. Price, Secretary-General, The Uranium Institute
*Monsieur M. Pecqueur, Directeur, adjoint à l'Administrateur Général Délégué, Commissariat à l'Energie Atomique (FRANCE)
Dr S. H. U. Bowie, FRS, The Institute of Geological Sciences (UK)
Prof. R. K. Dutkiewicz, Director, Energy Research Institute (SOUTH AFRICA)

Authors:
*Dr J. S. Foster (CANADA)
*Prof. P. L. Auer (USA)
Monsieur M. F. Duret, Atomic Energy of Canada Ltd
Mr R. M. Williams, Energy Mines and Resources (CANADA)

DEMAND AND CONSERVATION REVIEW BOARD

Chairman: His Excellency Dr Sumitro Djojohadikusumo, Minister of State for Research (INDONESIA)

Vice-Chairman: *Mr M. Celadon (TURKEY)

Members:
Mr R. Males, Electric Power Research Institute, California (USA)
*Dr K. Matsui, Institute of Energy Economics (JAPAN)
*Prof. Sir William Hawthorne, FRS (UK)
*Monsieur Jean-Romain Frisch, Electricité de France
*Dr H. du Moulin, Group Planning, Shell International (NETHERLANDS)
Dr N. A. White, Energy Consultant (UK)
Mr R. G. Chapman, Assistant General Manager, State Electricity Commission for Victoria (AUSTRALIA)

Authors:
*Dr R. J. Eden, Head of Energy Research Group, Cambridge University (UK)—Demand Report
*Mr L. Nevanlinna, Imatran Voima Osakeyhtioe (FINLAND)—Conservation Report
*Mr F. Kommonen, EKONO (FINLAND)—Conservation Report

The Panel was chaired by Dr A. A. Bestchinsky, Deputy Director, Electric Systems Network & Research (USSR).

WORLDWIDE PETROLEUM SUPPLY LIMITS

Ultimate resources and maximum annual production of conventional petroleum; possible resources of unconventional petroleum

Pierre Desprairies
Institut Français du Pétrole (IFP)

Worldwide petroleum supply limits

OUTLINE

I. In view of the difficulties of estimating ultimate worldwide resources in crude oil, and the limits of annual production capacity for this petroleum in the future, a Delphi-type poll was carried out in 1966-77 to obtain from experts throughout the world (oil companies, public authorities, individual experts) the basic data necessary for making such estimates. This was the first time such a poll had been carried out. 29 replies were received, out of 42 questionnaires sent out.

The poll covered conventional petroleum reserves (onshore and offshore at water depths of less than 200 meters), and also deep offshore and polar petroleum.

The most significant indications received were the following:

- ultimate worldwide conventional petroleum reserves are evaluated at around 250 to 300 GT (1 GT = 10^9 tonnes).
- the purpose served by continuing offshore prospection (45% of worldwide reserves) was confirmed.
- discovery costs are likely to increase substantially (more than double) between now and 1985/90. There will be a significantly smaller increase in the costs of developing deposits when discovered.
- the percentage of petroleum recovered from deposits will increase from 25% (1977) to 40% in the year 2,000.
- the increasing contribution of enhanced recovery (i.e. the expensive techniques required for such recovery) to the annual increase in reserves: 55% of gross increase in 2,000 are likely to represent re-evaluation of old deposits.
- the most disturbing point is the conclusion that the annual rate of growth of reserves is slowing down: in 2,000 it is likely to be at the most equal to total consumption in 1977 (3 GT).

II From among several curves of maximum possible technical production capacity, we offer two for serious study by those responsible. We feel that they give the most likely picture of the future with a production capacity ceiling of between 4 and 5 GT in 1990.

III 200 to 300 GT of unconventional petroleum (deep offshore and in the polar zones, heavy oils, enhanced recovery, oil shales, tar sands, synthetic oils) may be beginning to be exploited towards the end of the century at a price of $20 to 25 per barrel at 1976 prices. Its cost will confine it to specific utilization (transport, chemicals).

Unconventional petroleum seems likely to be essentially a supply source for the beginning of the XXIst century onwards. Vigorous action by Public Authorities (aid in developing processes) could accelerate its appearance and reduce its cost to around $15.

IV Numerous conditions would have to be satisfied for these technical production maxima to be achieved: financial aid to the companies, international agreements for the prospection and development of deposits in developing countries, to ensure that these deposits are not left unexploited, Government aid in improving prospection and production techniques. There is a considerable chance that for various reasons these maxima will not be achieved, or will be delayed. One of the principal problems is convincing public opinion of the existence and importance of the question at stake; without the support of the general public Governments will be unable to take the necessary political decisions to ensure that the investments are made in good time.

N.B 1 GT, one Gigatonne = 10^9 tonnes = 1,000,000,000 T
or 1,000 x 1 MT (one thousand million tonnes). 1 GT
≠≠ 7 x 10^9 barrels ≠≠ 44 EJ. 1 EJ = 10^{18} J.

INTRODUCTION

The Conservation Commission of the World Energy Conference, set up in Detroit in July 1974, has set itself the task of establishing the pattern of supply and demand for primary energy resources between now and the year 2020, and in particular, for petroleum,

- to inventory ultimate resources on the basis of the most authoritative expert opinion;

- to trace if possible a curve of maximum production available each year;

- and to list all possibilities of substitution by unconventional petroleum (tar sands, oil shales, etc.) evaluating their resources and exploitation possibilities.

The aim of this study is to give political leaders in the various Governments an idea of the maximum technical production capacities available, assuming optimum financial, statutory and political conditions.

Conventional petroleum is crude oil (together with condensates (1)) for which exploration and exploitation is currently being carried on with technology that is now considered classical, at a cost that is acceptable today. This in fact amounts to petroleum from onshore deposits and from offshore at less than 200 meters depth.

Unconventional petroleum requires for its exploration or exploitation techniques that are not yet fully developed, and whose profitability appears uncertain or insufficient today. Unconventional petroleum will require technological breakthroughs and is unlikely to appear in large quantities before 1990. It includes oil from deep offshore and the polar zones, most of the heavy oils and petroleum produced by enhanced recovery, tar sands, oil shales, and fuels of a synthetic nature derived from coal.

To evaluate ultimate recoverable reserves of conventional petroleum and to try to draw curves of maximum production capacity for conventional petroleum, the author initiated with the active support of the World Energy Conference General Secretariat a Delphi-type poll, to which some thirty replies were received from the most authoritative world experts: in oil companies, specialist public authorities, consultant firms and individual experts. This poll took place between September 1976 and April 1977.

(1) Very light oil produced at the same time as natural gas.

Part 1: Conventional petroleum resources (1)

Though the classifications introduced in the United States and Canada are extremely useful, this study has kept to the conventional system for petroleum classification, which only covers recoverable reserves and resources:

- proved reserves: these are discovered reserves that can be recovered in the existing state of technology and of the market. They are in deposits that are already equipped or in the course of being equipped.

- probable reserves: these are reserves that have been discovered and will probably be exploited given slightly better technical and economic conditions: extensions of existing deposits, small isolated deposits.

- possible resources (2): these remain to be discovered, but there is a reasonable degree of probability of their existence.

° ° °

1.1. One of the difficulties in evaluating ultimate recoverable reserves is the uncertainty of the recovery rate for petroleum in the ground (at present between 25 and 30%). This depends on the characteristics of the deposit, and on the production investment made; the latter itself depends on economic conditions. Another is the major uncertainty as to the size of undiscovered petroleum resources, which are often in zones of which little is at present known.

Proved reserves have not increased despite the quadrupling of the price in 1973, for various reasons of which the principal one is that most of the price increase has not been used to increase reserves. This, in OPEC countries, has taken the form of taxes used essentially for purposes other than oil exploration. In the industrialized countries the increased margins left to the companies have been largely absorbed by rises in costs. The sums invested in exploration or enhanced recovery have not yet had the time necessary to produce results.

(1) Petroleum reserves are already-discovered accumulations of oil; petroleum resources are those that have not yet been discovered, or the total of discovered and undiscovered quantities.

(2) the usual oil term is "possible reserves".

1.2. The Delphi poll

1.2.1. It seems that this is the first time a study of the Delphi type (from the name of the famous oracle in the ancient Greek city of Delphi) has been carried out to determine some presumptive but nevertheless important data for a long-term policy of worldwide petroleum resources: the amount of ultimately recoverable conventional petroleum resources and where they are, future costs of discovery and deposit development, future production costs, the relative importance of on- and offshore reserves, the future recovery rate of reserves in the ground, annual discovery rates, the relative importance in the future of newly-discovered deposits and re-evaluation of existing deposits. An original study on these essential points would have been a long operation; and its results would certainly have been less valuable than a direct poll of leading world experts. The questions involved concern the future. They can hardly be answered by calculation or extending existing curves; they call for a complex combination of knowledge, experience and judgment. Take for example the question of ultimate petroleum reserves. To determine the oil possibilities of an unknown or little-known region, the geologists employ complex analogical reasoning and probability calculations, which they do not always explain in detail, based on every aspect of their knowledge and their intuition; they work from well- or better-known regions, the presumed volume of the sediments in the zone to be explored, and the fragmentary geological knowledge in their possession. Such questions seemed very suitable for a Delphi poll, in which each expert is interrogated once, and then asked whether he wishes to maintain or modify his reply after having seen the other replies. The aim of this double question is to obtain concordant answers wherever possible.

42 questionnaires were sent out. 29 replies were received, including 2 from public authorities, 18 from oil companies, 9 from independent experts and consultancies; 20 answered both series of questionnaires. In addition, personal visits in the U.S.A., Canada and Europe helped to elucidate further and fill out certain replies. The absence of response from the socialist countries and the producer developing countries is to be regretted; they would have helped to widen, as was our object, the spread of opinion from competent experts. The questionnaire is given in an appendix. Its orginal version had 8 questions, and its final version 11, in an attempt to fix benchmarks for the costs and periods of time required to exploit the ultimate reserves.

The following replied to the questionnaire:

ATLANTIC RICHFIELD COMPANY (ARCO) (2)
THE AUSTRALIAN NATIONAL COMMITTEE OF THE W.E.C. (Miss J.C. MILLER)
Professor Dr. G. BISCHOFF, COLOGNE, WEST GERMANY
BRITISH PETROLEUM (B.P.) (2)
COMPAGNIE FRANCAISE DES PETROLES (C.F.P.) (2)
CONTINENTAL OIL COMPANY (CONOCO) (2)
DE GOLYER and McNAUGHTON (W.G. NANCARROW)
SOCIETE NATIONALE ELF AQUITAINE (Production) (2)
ENTE NAZIONALE IDROCARBURI (E.N.I.)
EXXON CORPORATION (2)
GULF ENERGY AND MINERALS COMPANY (2)
D.C. ION (2)
A.A. MEYERHOFF (2)
MOBIL OIL CORPORATION (2)
J.D. MOODY (2)
OCCIDENTAL PETROLEUM CORPORATION
PETER ODELL
PETROFINA S.A.
PHILLIPS PETROLEUM CORPORATION (2)
SHELL INTERNATIONAL PETROLEUM MAATSCHAPPIJ B.V. (2)
STANDARD OIL COMPANY OF CALIFORNIA (STANCAL) (2)
STANDARD OIL COMPANY OF INDIANA (AMOCO) (2)
SUN COMPANY (2)
TEXACO INC.
OWEN D. THOMAS (2)
UNITED STATES GEOLOGICAL SURVEY (U.S.G.S.) (2)
WEEKS NATURAL RESOURCES INC. (H.D. KLEMME) (2)

plus two replies from persons who asked their identities to be withheld, one of whom replied to both questionnaires (2).

We would like here to thank all these authorities, companies and experts most warmly for their help, and for their comments and advice which we have found extremely helpful.

The figure (2) indicates a reply to both questionnaires.

1.2.2. Results of the Delphi poll

The figures given below, and also the curves of maximum production derived from them, are an evaluation of technical maxima in respect of what nature and available technology will make it possible to produce within the limit of a cost price increasing up to $20 (1976)[°] per barrel in the year 2000, assuming that all conditions, and especially financial and political conditions, are favorable; they do not constitute production forecasts or predictions.

The following summarizes the essential data obtained from the Delphi poll:

1. Ultimate recoverable resources worldwide remaining to be produced as of 1977, supposing that the present recovery rate of 25% is raised to 40% towards the end of the century, have been estimated by all the 28 experts at 260 GT approximately, without counting deep offshore and the polar regions, which are still classified as unconventional petroleum; including these the estimate is 300 GT.

   These 260 GT include about 100 GT of proved and probable reserves already discovered, and 160 GT of reserves still to be discovered.

   If we limit our analysis of the replies to conventional petroleum (260 GT) we find three opinions among the experts:

   - a majority opinion (2/3 of the replies) estimating ultimate recoverable resources at 240 GT,

   - an optimistic opinion (25% of replies) of 350 GT,

   - a pessimistic opinion (10%) of 175 GT.

2. The 260 GT were considered by the 28 experts as likely to be divided as follows: U.S.A./Canada 11%; Western Europe 4.5%; U.R.S.S./China and the socialist countries 23%; Middle East and North Africa 42%; Africa south of the Sahara 4.5%; Latin America 9%; South and East Asia 6%.[°°]

   The opinions on deep offshore and the polar regions were pessimistic or uncertain, with an average of 13% of the 300 GT constituting ultimate resources.

(°) The concept of technical cost is explained in the definitions (p.1) for the questionnaire of the Delphi poll.

(°°) A speaker from the audience (Mr. Vedivaldi) expressed the desire for a more detailed breakdown of the answers by regions and countries. It is doubtful whether the experts would be prepared to commit themselves in such detail and would have answered a questionnaire so phrased.

3. The importance attributed to the Middle East/North Africa zone (essentially the Middle East) is considerable. The present recovery rate, probably lower than elsewhere, of these vast reserves suggest that the figure of 42% should be substantially increased. Note however that this percentage of 42% of ultimate resources is considerably smaller than that of the present figure of proved reserves (60%).

4. The reserves attributed to the socialist countries (23%) are considerable; however the extent of their national needs now and in the future suggests that they will export little or none of their output.

5. It is important to continue offshore prospection, since 45% of world reserves are presumed to be there.

6. The experts are relatively optimistic about the future cost of petroleum production. More than half of the petroleum remaining to be exploited could be produced at costs less than present selling prices ($12 in 1976); a third could be produced at around present production costs (less than $5).(°)

7. On the essential question of future discovery rates, most of the experts (2/3) are relatively optimistic for 1985, with gross annual discoveries (new fields plus re-evaluation of old fields) of about 4 GT per year, as compared with an average of 3 GT between 1950 and 1975. 1/3 only considered that the figure would be between 2 and 3 GT. On the contrary, expert views about 2000 are rather more pessimistic, with an average of yearly gross discoveries of 3.3 GT; a majority of experts even consider 3 GT as a maximum. Out of these gross discoveries, only 45% would constitute net discoveries of new deposits, with the other 55% accounted for by re-evaluation of old deposits, using enhanced recovery widely.

   We should remember that petroleum consumption in 1977 is approximately 3 GT. The experts estimate that at the end of the century discoveries will not provide for the renewal of reserves at the present level of consumption. The need to call massively on nuclear electricity and coal is thus clearly shown.

8. Between now and 1985/90 it is estimated that prospection efforts will have to be at least doubled to obtain the same oil discoveries.

   The reply is more optimistic on the costs of developing deposits at this same period (1985/90). Onshore, costs are are likely scarcely to exceed present development costs in the U.S.A. ($5,000/b/d.). Development in the conventional marine zones will probably remain at about its present level for difficult offshore conditions ($10,000/b/d).

(°) These are wellhead costs. During the poll, several experts verbally expressed the opinion that most of the ultimate resources could be produced at a technical cost of less than $12, and that $20 was a ceiling that would rarely be attained for conventional oil.

Deep offshore is likely to cost between one half as much again and twice as much as conventional offshore.

9. Enhanced recovery is expected to play an essential role towards the end of the century. At present, only 25 - 30% of petroleum in the ground is extracted; this figure will probably increase to 40% in about 2000: 45% in the industrialized countries, 42% in the socialist countries and 38% elsewhere.

10. Lastly, with regard to resources in natural gas, the experts considered that these will probably represent 83% of crude oil reserves in thermal equivalent. This proportion is higher than that generally agreed for proved reserves (63,000 Gm3 of natural gas compared with 90 GT of crude oil, or approximately 70%).

o o o

1.2.3. Comments on the estimates of world reserves given by the Delphi poll

- With 120 GT out of 257 (the average for all the experts) the Middle East/Africa area accounts for the largest single share of world reserves - almost half. It is also probably the area where enhanced recovery will give the most results.

- With 59 GT, or 23%, the socialist countries as a whole constitute the second largest group. The size of its present and potential internal market suggests that it will have little for export. But at all events the extent of its sedimentary area should render this group self-supporting.

- Conventional petroleum reserves are running out. The replies to the questionnaire are pessimistic on the rate of future discoveries (6.1.). Increased prospection costs (6.2.) and to a lesser degree development costs (10), and the higher share of production obtained by enhanced recovery (9) and of discoveries constituted by re-evaluation (6.1) confirm this opinion, which concurs with that expressed over a number of years by numerous experts and companies (J.D. MOODY, EXXON, BP, H.R. WARMAN, etc.) and recently by H.D. KLEMME (March 1977). This author cites (a) the essential contribution of the "giant" fields (more than 500 x 10^6 barrels, or 68 x 10^6 tonnes of petroleum, and 3.5 TCF, or 100 x 10^9 cubic meters of natural gas) to worldwide oil demand, (°) and notes that while between 1970 and 1975

(°) at present 70%

giant hydrocarbon fields were still being discovered 72% of these were gasfields (as against only 10% of the giant fields discovered around 1955), and (b) emphasizes that between 1970 and 1975 only 8.5 billion barrels of oil were discovered each year (1.2 GT) in the form of giant fields. (The author includes in these 8.5 billion barrels deposits discovered before 1970 but which only became classifiable as giant after 1970 - these account for 30% of the total.)

If we suppose that the giant oil fields contain on average 75% of total reserves, total discoveries of new oilfields would, according to Mr. Klemme, have provided around 1.5 GT per year between 1970 and 1975, whereas between these same dates average annual world consumption was 2.65 GT. This figure of 1.5 GT agrees with that given for the same period by M.H. Warman (Revue Française de l'Energie, November 1975).

During the discussion, one speaker (Mr. Pennel) pointed out that of the 30,000 existing oil deposits, only 300, or 1%, contained 80% of reserves, and that whereas the increase in prices would perhaps multiply proportionally the number of deposits discovered, it would not increase the proportion of large to small deposits, or increase available reserves proportionally. This remark was supported by another speaker (Mr. Martinez). A third member of the panel (Mr. Attiga) however said that the rise in prices was stimulating the exploration effort (noticeable since 1973). The Vice-Chairman of the Round Table (Mr. Kafescioglu) suggested that national interests might make it desirable to produce oil at costs exceeding world prices.

These are the main results of the study, which we believe to be the first of its kind to be undertaken. They concur with the figures generally accepted with regard to ultimate resources of conventional petroleum. They confirm the importance in the future of the Middle East, though this will be less than its importance today. The usefulness of offshore prospection is also confirmed. The costs of future production and development should remain reasonable, though discovery costs are likely to increase substantially. The percentage of petroleum recovered from the deposits should increase by more than half by the end of the century; enhanced recovery will by this time probably play an essential part in the annual increase in reserves. The most disturbing point is the total amount of discoveries made annually, which all the experts expect to decline.

Part 2. The technical limits of annual production of conventional petroleum

Some curves have been drawn to evaluate the maximum technical production capacity per year based on the ultimate reserves as indicated by the experts. They are derived from curves of cumulative discovery by zone assuming that petroleum discoveries in the world's oil-bearing zones as a whole will follow the pattern of the United States, which is a country of varied geology which has been extensively explored, and which is now entering a period of declining discovery progression, after a phase of exponential followed by linear growth. The United States have been considered the best model for rapid and effective exploration in replying to the question, which was how to arrive at maximum world production capacity. The curve chosen to express cumulative discoveries, sometimes called "logistics", assumes the a priori reasonable hypothesis that the "tail off" period on the cumulative-discoveries curve will be symmetrical with the "take off" period, and that the middle of the curve will correspond to the period of linear growth of cumulative discoveries.

For each of the six major oil-bearing areas (North America, Latin America, Western Europe, the Socialist countries, the Far East, the Middle East/Africa), and also for the world as a whole, three curves have been drawn based on the American pattern firstly of cumulative discoveries and secondly of annual discoveries, taking three hypotheses of ultimate resources: a) low of 175 GT, b) medium of 240 GT, and c) high of 350 GT. For each of these three hypotheses, three depletion rates have been posited: a) the 1975 rate (10 years for North America, 23 for Latin America, 15 for Western Europe - the probable cruising rate -26 years for the Socialist countries, 28 for the Far East, and 54 years for the Middle East and Africa). The corresponding rate for the world as a whole would be 37 years (i.e. on January 1st of each year a decision would be taken to produce 1/37th of the reserves existing at that date).

b) a faster rate, described as "medium": 10 years for North America, 15 for Latin America, 15 for Western Europe, 20 for the Socialist countries, 20 for the Far East, 35 for the Middle East. The world rate would then be 25 years.

c) a rate described as "high": the same as above, but 25 years for the Middle East/Africa area. This rate is technically possible, but unlikely in practice. It has been taken to demonstrate the upper technical limit of world production capacities. The world rate would then be 20 years.

The varying degree of probability of each of these curves is open to discussion(°). The most optimistic ones are the least probable; some of these might pose technical problems of developing capacity. It is also possible to argue that up to around 1985 the growth in world production capacity will be limited by world demand, which it exceeds at the moment. The two most likely curves, D I and D II, are based on the reserves estimated by the central group of experts (238 GT) and "medium" 1975 depletion rates (with 35 years for the Middle East/Africa region)(°°). They suggest a maximum technical capacity worldwide of 4 to 5 GT/year in about 1990, falling back to 3.4/4 GT/year in 2000 and 2.5 GT/year in 2020. It is these two curves that we propose as a basis for consideration by political and industrial leaders.

World consumption, it should be remembered, was about 3 GT in 1977.

In graph F, we have tried to show the preponderant influence of the Middle East/Africa region on world production maxima.

To this end, using a figure of 239 GT for ultimate recoverable resources, we have assumed a "medium" depletion rate for all oil-bearing regions with the exception of the Middle East/Africa, for which we have posited depletion rates at the 1975 level (54 years), of 35 years, and of 25 years, which last two rates are technically possible. These three hypotheses show that during the critical years 1985-1990, the depletion rate adopted for this region alone can cause the maximum technical production capacity worldwide to vary by 1.2/1.3 GT/year, without reducing world production capacity in 2020. These figures corroborate recent studies, which lead to the same conclusion. They are not surprising if we remember that the resources of the Middle East/Africa zone (6/7ths of which are in the Middle East) represent nearly 50% of all the conventional petroleum resources in the world. Note however that for the whole Middle East/North Africa area the share of 42% of ultimate conventional resources assigned to it by the experts is less than its current share of 60% of proved resources.

The share of resources and production capacities held by the countries with planned economies is high (23% of world resources) but it is presumed that between now and 2020 this group of countries will be self-sufficient, or will export limited quantities only.

(°) See these 9 curves in figures C, D and E of appendix II.

(°°) These also correspond well to the forecasts of annual discoveries in 1985 and 2000 made by the Delphi Survey experts (see fig. B in appendix II).

A comparison has been made (figures H, I, J) with the "optimistic" hypothesis (C 1) of the recent WAES (Workshop on Alternative Energy Supplies) report of May 1977, completed to include the production of the Socialist countries. It suggested a situation fairly similar to that of the most "optimistic" of the curves drawn on the basis of resources of 240 GT. On the other hand, application to the production capacity picture considered as most probable (curve D II) of the ceiling of annual discoveries used in the same WAES optimistic hypothesis C 1 completely changes production prospects, with a maximum of only 4 GT, reached in 1980. This confirms the importance of the factor of annual discoveries in fixing future production capacities.

The years 1985 to 1995 seem critical. During this period production capacities are likely to level off in almost every hypothesis. If the demand for petroleum continues to grow at that time, i.e.if there is not sufficient availablity of coal and nuclear electricity, there will probably be a shortage and oil prices are likely to rocket uncontrollably.

Part 3. Unconventional petroleum

These reserves are inadequately evaluated, in view of their low economic interest at the present time. They do not offer sufficient profitability under present economic conditions and the necessary technology either does not yet exist or has not been fully developed.

3.1 Oil shales. The reserves in the ground are at present evaluated at 400 GT, only 30 GT of which are exploitable with current technology. It is likely that when production starts this will be in the United States (Colorado). The ecological problems (water, spoil matter) are considerable. Investment costs are of the order of $20,000 per b/d. Production costs are at present estimated at between $20 and 25 (1976) per barrel, with an interest rate of 15%. If the hopes raised by the Garrett process (semi-in-situ combustion, production costs of $8 - 11/bl, much less acute ecological problems) are not confirmed, shales are not likely to play a very important role until the end of the century. Substantial Goverment aid will in any case be necessary for development of the processes.

3.2 Tar sands and heavy oils. World reserves are estimated at 300 (or 350) GT 5-10% only of which are exploitable on the surface. It is in Alberta, Canada (120 GT in the ground), with the GCOS plant (50,000 bbl/d in 1967) and soon with Syncrude (125,000 bbl/d in 1978) that surface exploitation work is most advanced. Venezuela will then probably begin large-scale exploitation of similar and sometimes lighter hydrocarbon (Orinoco Belt, 100 GT in the ground). In their paper to the Xth World Energy Conference (N° 1.4.13) Messrs Gutierrez, Parra and Gonzalez estimated that reserves in the ground could exceed 300 GT. The recovery factor they suggested was 10%, giving 15 GT of the 150 GT in the ground that are sure.

The cost price ex-tax for asphaltic crudes is estimated at $25 (1976) per bbl approximately with interest at 15%, or $15 with 9% interest.

3.3 Enhanced recovery can increase the present average recovery rate of 25-30% - the current average - up to 45-50%.

In the course of a discussion on a remark by Dr. Kirby, the author said that the figure of 25-30% (usually 25%) had been given to him verbally by several experts as the closest probable approximation. The highest figure for current recovery (45%) is that given by the Russians, who inject water systematically. Mr. Pennell drew attention to the imprecision of the figures currently available, and to the need for an international study of the problem. Mr. Kafescioglu, taking as an example the use of enhanced recovery on deposits in Turkey, showed that even on this smaller scale the additional supplies for the world market were far from negligible. Mr. Martinez drew attention to the large reserves of heavy oil available in Venezuela.

Two recent American studies (Lewin 1976, NPC 1976), based on detailed analyses of existing deposits, point up both the hopes generated by this process (an 85% increase on current reserves in the United States, and a 50% increase in production over 1976), and also its handicaps: heavy investment, uncertain technical results which may vary by a factor of one to three or even four, high production costs (production will increase most significantly at between $8 and $15 per barrel), long pay-off periods (10 - 12 years) and unattractive cost-effectiveness in view of the investment involved and the technological risk.

A theoretical calculation applied to the 650 GT in the ground remaining to be produced in 1977 gives an increase of 100 GT in recoverable reserves, if the recovery rate increases from 25% to 40%.

3.4 Deep offshore and Arctic zones. The experts of the Delphi poll in general showed little enthusiasm for deep offshore (over 200 meters). Somewhat less than half gave it 7 to 15% of world resources, a similar proportion considered these reserves as small or negligible, and only 7% though that around one third of resources may be found there. One specialist in this problem (Weeks) estimated that deep offshore - the continental slope, or the transition between the continental shelf and the abyssal plains - may contain 16% of world petroleum resources.° The geological uncertainties (presence of oil, reservoirs, traps, overburden) can only be resolved by drilling. The technology is being progressively developed, but little is yet known as to exploration and exploitation costs.

The polar zones pose problems both of production and transport. It seems hardly likely that they will be exploited before the end of the century.

3.5 Hydrocarbons derived from coal, especially by the new extraction processes now being experimented with using solvents in the presence of hydrogen, could create interest because the yield is fairly high (70 to 75% or about 2 tonnes of coal for 1 Tep of crude oil). However the investment costs are heavy ($20,000 per b/d) as are production costs ($20 - 25 per barrel, with 15% interest on capital invested). The coal would also have to be cheap. This, together with high-calorific-value synthetic gas, is probably one of the ways coal will be used in the future.

°i.e. about 50 GT

In reply to Mr. W. K. Davis's question, the author said that he had not studied problems of the development or financing of coal reserves linked to this production of synthetic oil because the problem appeared to him to be in the field covered by the "coal" resources group. Mr. McCormick pointed out the legal obstacles and those linked to the environment that are currently opposing the development of coal resources in the United States. In response to a comment by Mr. Attiga that national energy producing countries -- coal or oil -- could refuse to export it (such as the recent refusal to export oil from Alaska to Japan), the author of the report said that the energy market is international and that any reduction in demand on this market is the equivalent of a supply.

Fuels of vegetable origin (methanol) are likely to fall into the category of expensive products, because of the costs of agricultural production, collecting and transformation in large plants of a low-grade raw material.

3.6 The unconventional oils will above all be exploited in the XXlst century, succeeding conventional petroleum in its specific applications (transport, petrochemicals). Government aid will be necessary for its production. However one should note that once the technology has ceased to involve risk, and the companies can be content with normal industrial profitability rates (8 to 9%), the cost of these oils could be around $15 per barrel and thus not greatly exceed the current selling price of crude. They may well thus constitute in ten to fifteen years a useful stabilizing element, keeping down prices on the world oil market, provided Governments are prepared to give active support to developing the technology necessary for their production.

Mr. William Sande underlined the fact that all investments in unconventional energy production units of international size (tar sands, gasification of coal, nuclear power plants) cost from $2 to 3 billion per unit, require the mobilization of extensive construction manpower, and take 5 to 7 years to attain achievement.

Part 4. The conditions for achieving maximum production capacities

The only question that is of legitimate question to politicians and the general public is: "Will there be enough oil to satisfy world demand in the next 30 or 40 years?" The reply is: "Yes, if we include oil from all sources; no, if we mean the conventional petroleum produced cheaply or relatively cheaply that we use today for any form of energy." There is probably not enough conventional petroleum at relatively short term to satisfy all the demand that, at its present growth rate, seems likely to exceed the forecast maxima of 4 - 5 GT/year in about 1990. There will be enough expensive petroleum to satisfy a limited level of consumption in applications specific to petroleum (transport and the chemical industry), with the other needs (°) being covered by cheaper forms of energy (coal and nuclear).

However, if these maxima of 4 - 5 GT/year as forecast are themselves to become reality, the necessary investment will have to have been made at the right time and in the right places; otherwise the world economy may be upset by acute crises in supply. A number of conditions, financial, economic, technological and political will have to be fulfilled.

4.1 Sufficient financial resources (in capital, loans, and own finance) will have to be available to the companies, both private and public. These are very far from being able to call on the sums necessary to finance the estimated costs of future production.

Profound changes in present ideas of price, margins and the role of petroleum as a tax source must be progressively envisaged. Similar changes will also be necessary in relations between Governments and the companies, in both the private and public sectors, with a view to creating or restoring a spirit of cooperation which is at present lacking in many countries, especially certain industrial countries.

4.2 Maximum world capacity of production in the next two or three decades will be determined by the rate of new deposit discovery and by that of bringing into operation methods of enhanced recovery. The maxima indicated in this report can only be attained if the investment effort in the fields of exploration and recovery is applied in all the oil regions of the world in rough proportion to the chances of finding resources there. The maxima will have less chance of being achieved if money invested is directed preferentially, as it is at present, to less promising zones leaving aside the developing countries. It is in the interest of all mankind

° about half of present petroleum needs

that the limited means of exploration and development at its disposal should be used as efficiently as possible. Even if the attempts made over the last two years to bridge the gap between North and South have had disappointing results, it is only by travelling the rocky road of agreement between us that we may hope to achieve, by the end of the century, production levels that will satisfy the world's needs.

Unless the oil-producing developing countries themselves make a serious effort in oil investment, international agreements will be necessary to ensure that the money required to mobilize resources is not invested mainly outside the developing countries, where 60% of world resources, and the least expensive ones at that, are to be found. This would also seem to be in the interests of countries within these zones that do not have the means or the desire to make such investments themselves.

As world suppliers, the Middle East especially and also Africa and Latin America will retain pride of place over the long term.

4.3 Substantial Government aid will be necessary for the appropriate prospection and production <u>technologies</u> to be made available. Training specialized <u>labor</u> in the new production techniques may constitute a second bottleneck.

4.4 There are thus evident risks of delay in the achievement of maximum production capacities, and also risks that they will not be fully achieved at all: increased costs due to the exhaustion of natural resources; a general unwillingness to make long-term investments in times of inflation; very uneven participation by the oil-producing countries in the course of development in investment for petroleum exploration and production; the tendency of the oil companies only to invest in the industrial countries, even though these are poor in oil reserves. The favorable counterpart for such delays, the considerable prolongation of production life into the next century, is unfortunately not of great interest, since the problem of changeover from conventional petroleum to coal and nuclear energy on the one hand and to unconventional oils on the other is one that must be faced in the next fifteen to twenty years.

4.5 While the limits imposed by nature and technology on the exploitation of resources seem increasingly real and restrictive, those in the field of politics are of even more decisive importance. We should also remember that when production capacity is installed, its maximum working level and the period and rapidity of its decline constitute political decisions for which sovereign states have full jurisdiction.

4.6 Among all the obstacles separating the technical possibilities of petroleum production from their effective realization, the most difficult to surmount is failure to believe that they exist. The scale of investment necessary requires that the general public should be convinced that the problems confronting it are real, and should be willing to help solve them. The amount of information published on the subject in the last three years, and the present abundant supplies (even if people are beginning to recognize that these are temporary and largely artificial) have created an almost universal scepticism with regard to the problem of petroleum resources. What can be said now to make people believe in it? Will some event that interferes with their daily lives have to occur for them to be convinced? Should we, after all, hope that such an event does not come too late?

GENERAL OBSERVATIONS

Some of the points taken up during the Round Table Discussion on September 21 1977 have to do with several problems covered in the report.

1. The chances of technical innovation. A speaker from the audience (Mr. Igor Visusky) emphasized the technical advances made in recent years in the field of exploration, and asked whether forseeable advances had been taken into consideration in making forecasts of discoveries. The author of the report replied that they had indeed been taken into account. However, technical progress in exploration cannot multiply the number of giant fields, which contain most of the reserves, and it is not at present possible to predict a technological revolution that would appreciably increase the discovery rate in the future. This opinion was shared by another member of the panel (Mr. Pennell).

2. The price of oil. Concerning the recovery rate, which is at present estimated at 25%, Mr. Attiga felt that this rate could be increased at less cost in the exporting than in the industrialized countries. This improvement, which it is the responsibility of the producing countries to make, is possible only with higher prices. Taking into account a replacement cost of $20, this member of the audience felt that the exporting countries are at present losing $8 per barrel. A price rise would result in the Middle East reserves being increased by 50 to 100%. Mr. Attiga also pointed out that the 1973/74 price rise from $2 to $12 per barrel, corresponding to only 2% of the GNP of the industrialized countries, cannot be held responsible for the world-wide inflation.

   The author of the report said that a rise such as the one mentioned would be difficult to justify in economic terms.

One the one hand, it is not very likely that the cost of producing conventional oil in 1985/90 will reach or go beyond $15 or $16 (1976). On the other hand, if oil attains production costs of $15 to $20 towards the end of the century, it will be replaced in all its uses other than specific ones (land and air transportation, petrochemicals), i.e. for 60% of its present uses, by coal and nuclear heat produced at $7 or $8 for the thermal equivalent of a barrel of oil. Therefore, it cannot be considered that $15 to $20 represents the replacement price in the future for oil used in 1977.

3. Financing exploration and development. The author of the report emphasized the importance, in his opinion, of the problem mentioned on page 17 of the executive summary if the production maxima of 4 to 5 GT/year are effectively to be attained. If the exporting countries, which have 60% of world resources, do not themselves make the effort required for prospection and development, companies in the industrialized countries will not begin large-scale investment in developing countries unless they are guaranteed, in the event of success, a contractual remuneration on a level comparable with that which they would obtain in the industrialized countries ($1 to $2 per barrel, whereas the remuneration proposed in the developing countries is 25 to 40 cents); and this remuneration will have to be guaranteed by an international mechanism ensuring that the contract is properly carried out under the terms agreed on. The useful assistance that the World Bank can provide to governments in some cases, such as to the Indian Government, for the start-up of production from the Bombay High field ($150 million), is extremely precious, but the resources currently available to the World Bank are not on the same scale as the problem.

   Mr. Attiga emphasized at this point that the development of energy resources in developing countries such as India will require considerable assistance from industrialized countries in the form of loans at advantageous rates, outlets for their industrial production, and low-cost transfer of technology. The richest OPEC countries - three or four in number - are already helping with this development to a considerable degree, but their contribution can only be financial and they can only provide a part of the necessary funds.

4. Should not future production-capacity curves have been plotted on price hypotheses and not on hypotheses of oil-production costs?

   In the course of the discussion, Mr. Attiga pointed out that the production forecast curves should have been based on the selling price, which will be the factor governing the investment rate, rather than on the production cost.

This would indeed have been preferable, but it would have entailed another report which we did not feel it possible to undertake on solid bases. Various hypotheses would have had to be chosen for production costs, production volumes of rival energies (nuclear electricity and coal), market price and especially taxes, and they would have had to be made on an evolving basis between now and 2020 in order to estimate the total revenues available for producing companies and the volume of investments. It is probable that this would have flattened out the curves and prolonged oil production over time, as was suggested by Mr. W.N. Sande in an observation received after the Conference. However, if the investments necessary for the production of sufficient amounts of nuclear electricity, coal and unconventional oil are not made in due time, oil prices will undergo a considerable rise, and the result of this will probably be that the margins available for investment are increased and that discoveries and the production of reserves are speeded up: in other words, the technical maxima for production capacities will be attained. The report presented in Istanbul did not attempt to predict the future but rather, with the help of the international experts, simply to estimate this technical maximum, i.e. the annual production level that it appears difficult to exceed in any hypothesis in the present and future state of technology. It is true that if the effective price level - or more precisely the margins at the disposal of the producing companies - is insufficient, these maxima will not be attained. Unfortunately, this will probably be the case, as is emphasied in Part 4 of the report.

APPENDICES

I. Ultimate recoverable resources:
Results of the Delphi poll.
(table)

II. Maximum production capacity.
(curves and tables A-K)

III. Questionnaire (N° 2).
Attached letter and note.

IV. Comments on Part 3. Unconventional petroleum
by Messrs W.A. Roberts, Executive Vice President,
Phillips Petroleum Company (plus 2 figures) and
W.N. Sande, Executive Vice President, Syncrude
Canada Ltd.

ULTIMATE RECOVERABLE RESOURCES OF OIL

G.t.

	U.S.S.R. and Eastern Europe	China	United States and Canada	Middle East and North Africa	Africa South of Sahara	Western Europe	Latin America	Japan - Australia New-Zealand	East and South Asia	Conventional Oil	Deep Offshore	Polar Areas	Total
1*	(20)	(5)	(10)	(57.5)	(3.5)	(7.5)	(8)	(5)		(116.5)	(2.5)	(7.5)	(126.5)
2	20.2	7.1	32	80.3	6.3	14	7.9	5.5		173.3	0	0	173.3
3	68.5	9.6	6.2	54.8	2.7	6.8	18.5	2.7	5.5	175.3	5.5		180.8
$\bar{X}$	52.7		19.1	67.6	4.5	10.4	13.2	6.8		174.3	2.8		177.1
4	41	11	26	85	11	22	16	11		224	ε	2	226
5	60		27	76	8	10	17	16		214	13		227
6	40.3		29.1	103	13.4	9	17.9	11.2		224	7 to 30		231
7	46		25	103	12	6	22	22		235	included		235
8	57.1		28.6	87.7	9.1	10.8	26.1	16		235.3	—	2.7	238
9	37.5	5	17.5	112.5	7.5	7.5	12.5	5	20	225	20	5	250
10	41+		41	82+	5-	5+	26	10		210	46		256
11	36.4	7.8	15.6	156	10.4	10.4	13	2.6	7.8	260	0	0	260
12	37.5	19.7	18.4	77.7	7.3	15.2	36.9	2.6	12.5	227.8	34.4	1.3	263.5
13	69	14	25	90	9	10	13	2	11	243	17	5	265
14										265.5			265.5
15	74		28	119	8	16	15	10		270	p.m.		270
16	44.4	16	18.7	134.3	8.9	8.7	12.7	2.9	6.5	253.1	24.4	-	277.5
17	59		20	110	4	7	34	6		240	50		290
18	45		45	90	9	18	30	15		252	48		300
19/20/21													<300
$\bar{X}$	54.5		26.1	101.9	8.8	11.1	20.9	13.6		238.6	21.2		257.0
s	12.7		8.5	23	2.5	4.9	8.4	5.1		18.4	18.9		22.6
$\bar{X}/s$	4.3		3.1	4.4	3.5	2.3	2.5	2.7		12.9	1.1		11.4
22													300< <500
23	52.5	24.5	42	122.5	14	14	35	21		325.5	24.5		350
24	80.5	15.75	43.75	118.3	12.25	14.7	31.5	15.75	10.5	343	7		350
25	40	25	50	100	20	5	55	30		325	150	25	500
26	75		25	90	20	15	20	25		270	180	50	500
27	60	5	35	300	40	10	20	5	5	470	15	15	500
28													550< <950
$\bar{X}$	75.7		39.2	146.2	21.3	11.7	32.3	22.5		346.7	93.3		435.7
$\bar{\bar{X}}$	59.4		28.5	109.1	11.3	11.2	22.9	15.1		257.3	38.7		302.4
$\bar{s}$	17		11.1	49.3	7.9	4.6	11.1	7.1		63.9	60.8		104.8
$\bar{\bar{X}}/\bar{s}$	3.5		2.6	2.2	1.4	2.5	2.05	2.1		4	0.6		2.9

* Includes future discoveries to year 2000 - not ultimate.

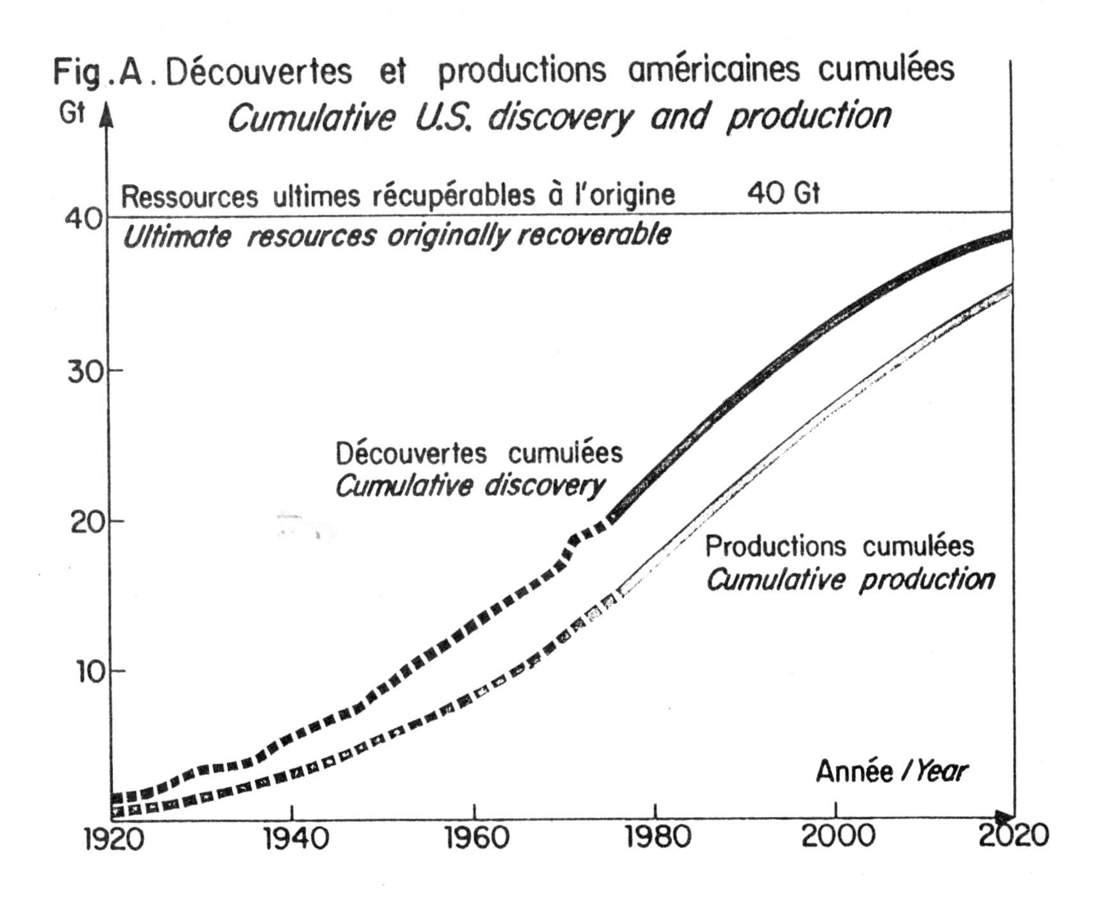
Fig.A. Découvertes et productions américaines cumulées
Cumulative U.S. discovery and production
Gt
Ressources ultimes récupérables à l'origine 40 Gt
Ultimate resources originally recoverable
40
30
20
10
Découvertes cumulées
Cumulative discovery
Productions cumulées
Cumulative production
Année / Year
1920
1940
1960
1980
2000
2020

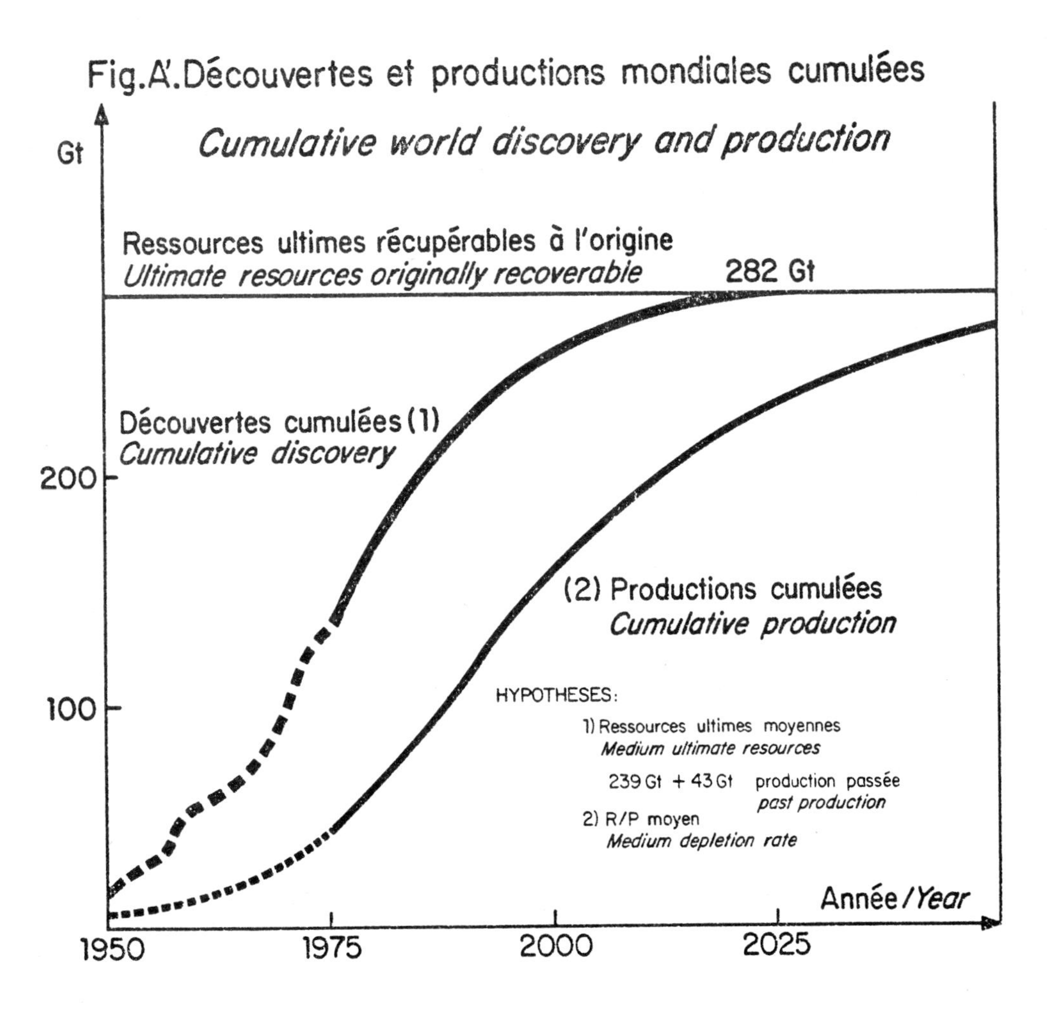
Fig.A'. Découvertes et productions mondiales cumulées
Cumulative world discovery and production
Gt
Ressources ultimes récupérables à l'origine
Ultimate resources originally recoverable 282 Gt
Découvertes cumulées (1)
Cumulative discovery
200
(2) Productions cumulées
Cumulative production
100
HYPOTHESES:
1) Ressources ultimes moyennes
Medium ultimate resources
239 Gt + 43 Gt production passée
past production
2) R/P moyen
Medium depletion rate
Année / Year
1950
1975
2000
2025

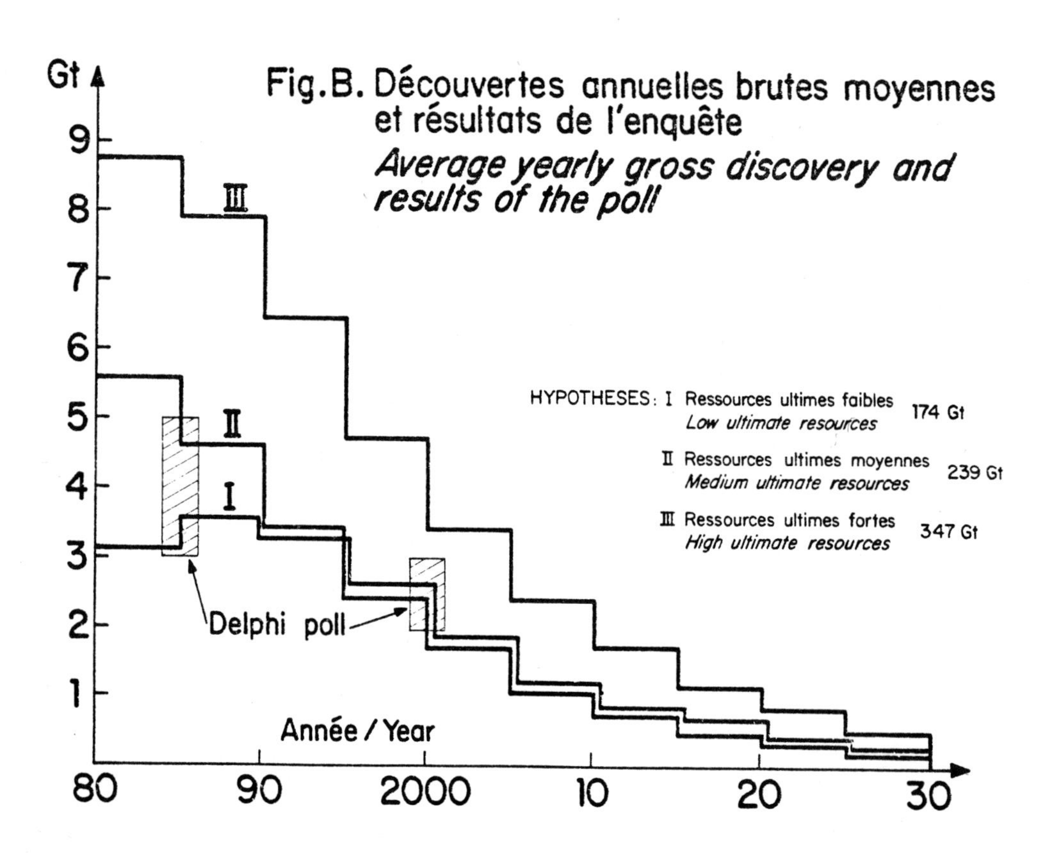

Gt
9
8
7
6
5
4
3
2
1
Fig.B. Découvertes annuelles brutes moyennes et résultats de l'enquête
Average yearly gross discovery and results of the poll
III
II
I
HYPOTHESES: I Ressources ultimes faibles
Low ultimate resources
174 Gt
II Ressources ultimes moyennes
Medium ultimate resources
239 Gt
III Ressources ultimes fortes
High ultimate resources
347 Gt
Delphi poll
Année / Year
80
90
2000
10
20
30

Fig.C. Capacités maximales de production
Ressources ultimes récupérables faibles: 174 Gt

Maximum production capacity
Low ultimate recoverable resources: 174 Gt

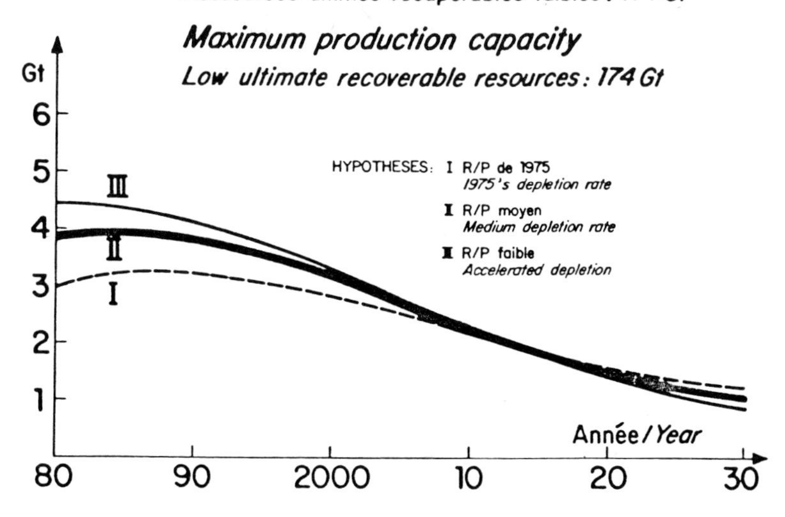

Fig.D. Capacités maximales de production
Ressources ultimes récupérables moyennes: 239 Gt

Maximum production capacity
Medium ultimate recoverable resources: 239 Gt

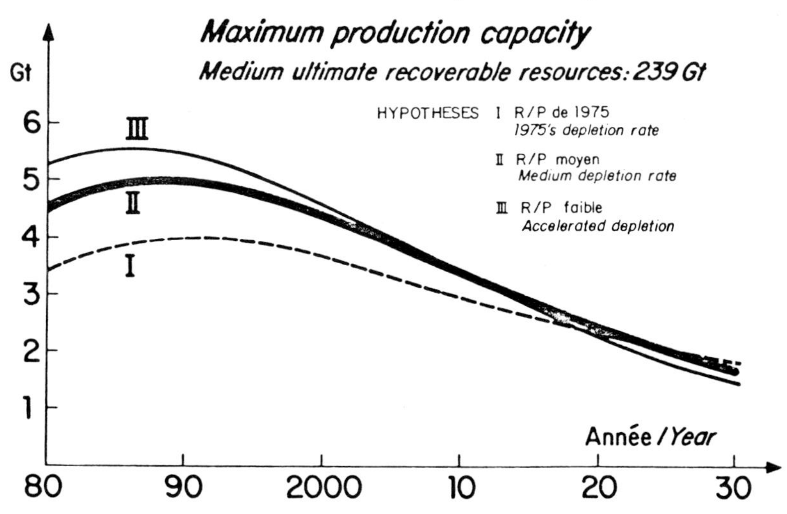

Fig.E. Capacités maximales de production
Ressources ultimes récupérables fortes: 347 Gt

Maximum production capacity
High ultimate recoverable resources: 347 Gt

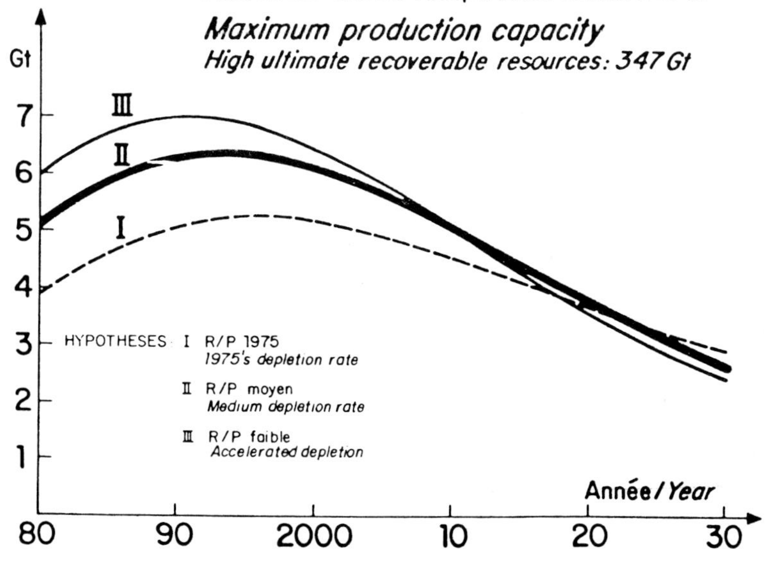

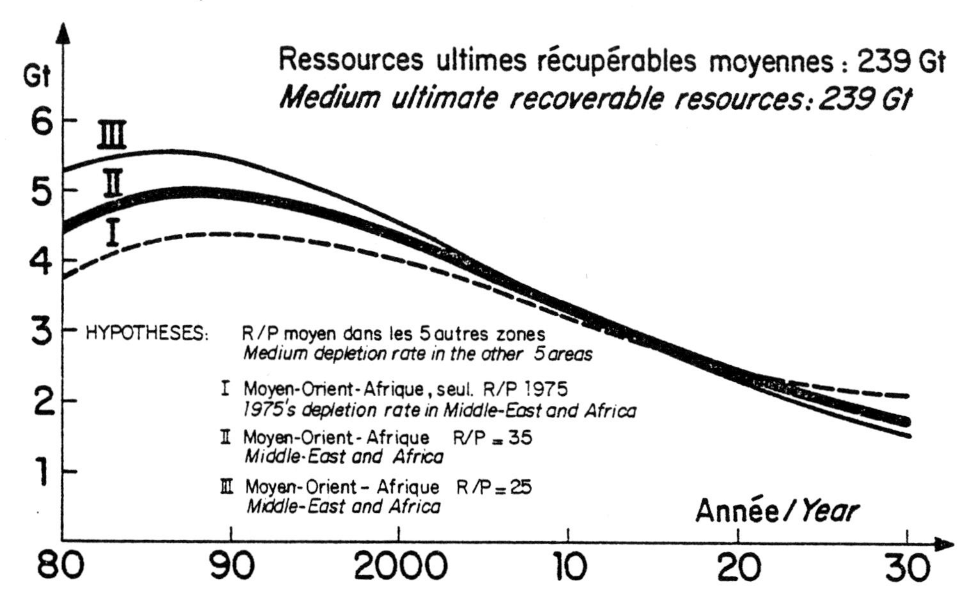
Fig.F. Importance de la région Moyen-Orient – Afrique
Impact of the Middle East – Africa zone
Gt
Ressources ultimes récupérables moyennes : 239 Gt
Medium ultimate recoverable resources: 239 Gt
6
5
4
3
2
1
III
II
I
HYPOTHESES:
R/P moyen dans les 5 autres zones
Medium depletion rate in the other 5 areas
I Moyen-Orient-Afrique, seul. R/P 1975
1975's depletion rate in Middle-East and Africa
II Moyen-Orient-Afrique R/P = 35
Middle-East and Africa
III Moyen-Orient-Afrique R/P = 25
Middle-East and Africa
Année / Year
80
90
2000
10
20
30

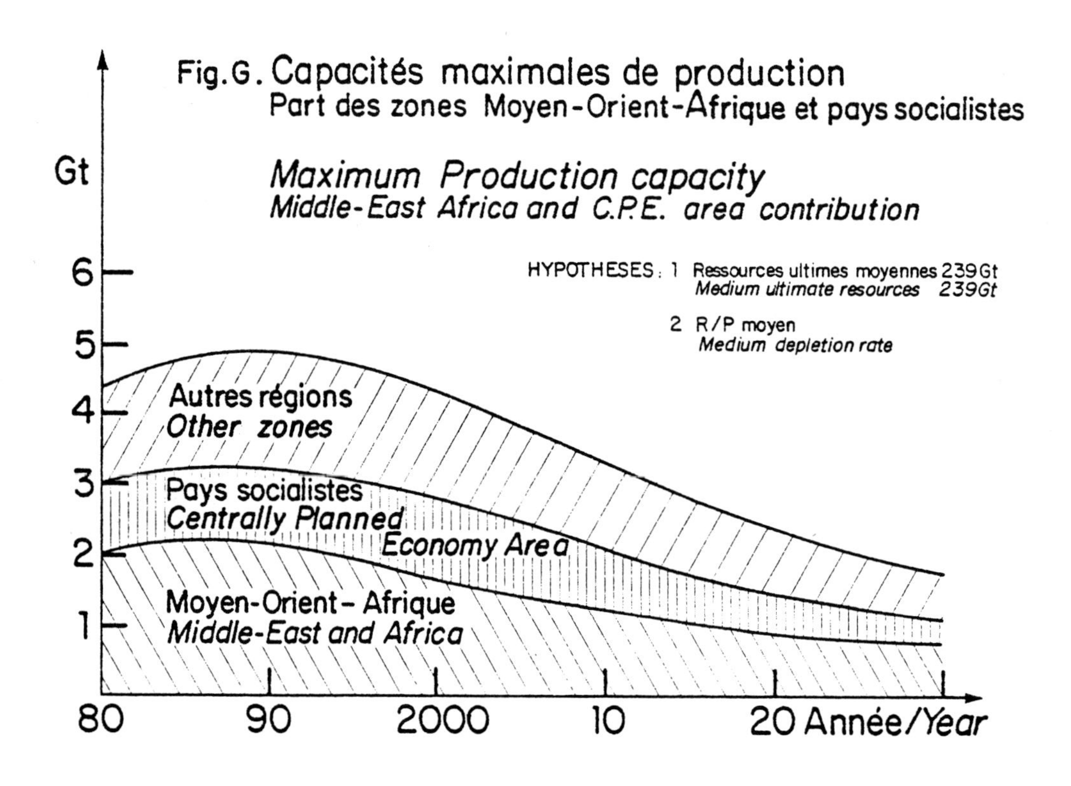
Fig.G. Capacités maximales de production
Part des zones Moyen-Orient-Afrique et pays socialistes
Maximum Production capacity
Middle-East Africa and C.P.E. area contribution
Gt
HYPOTHESES: 1 Ressources ultimes moyennes 239 Gt
Medium ultimate resources 239 Gt
2 R/P moyen
Medium depletion rate
6
5
4
3
2
1
Autres régions
Other zones
Pays socialistes
Centrally Planned Economy Area
Moyen-Orient - Afrique
Middle-East and Africa
80
90
2000
10
20 Année / Year

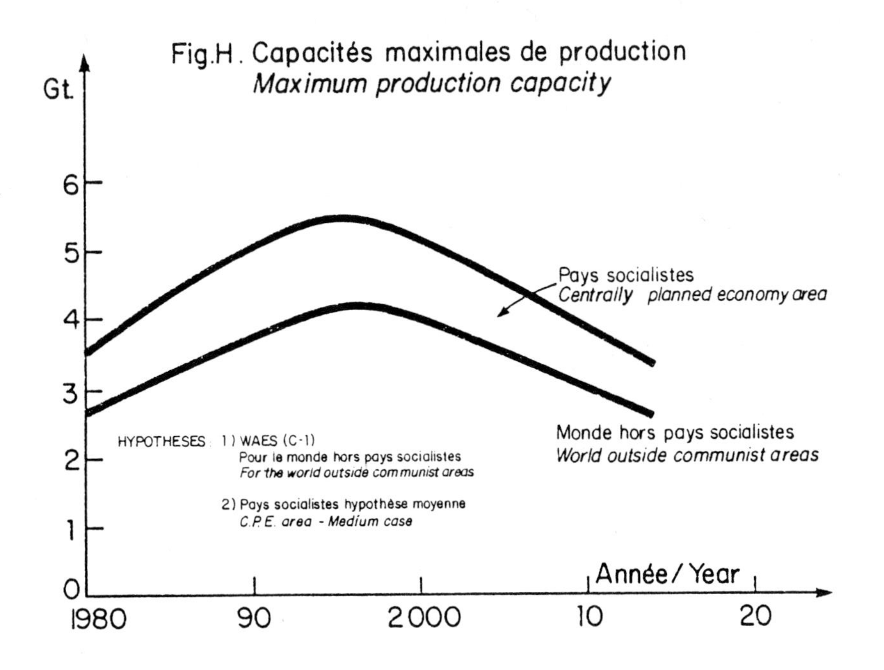

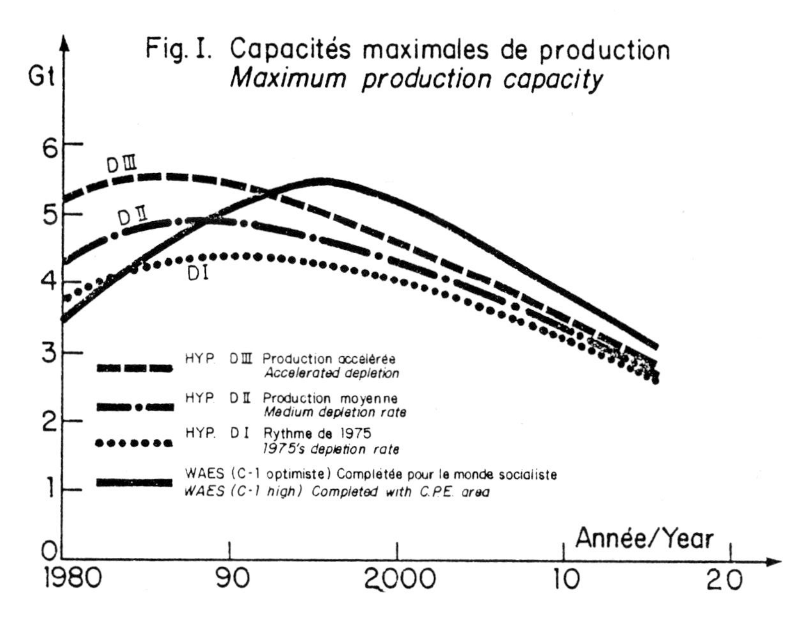

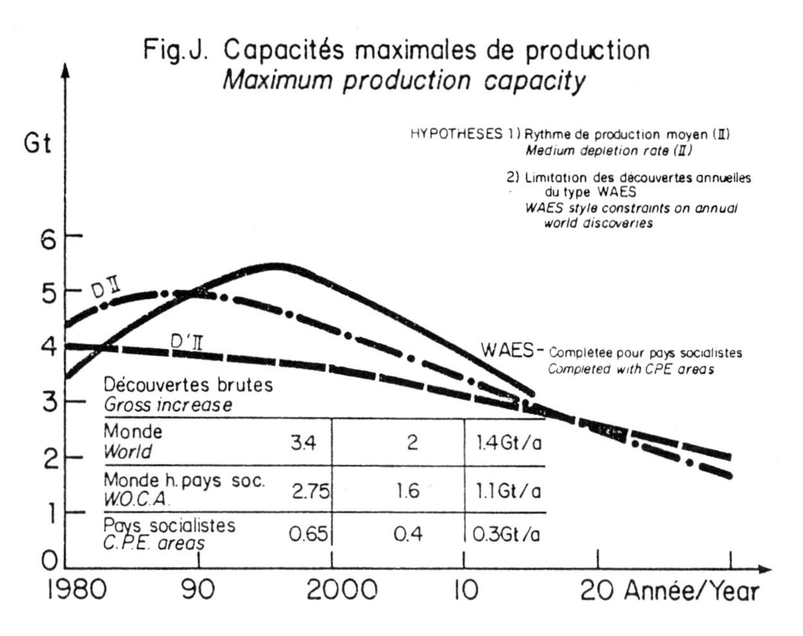

Découvertes brutes *Gross increase*			
Monde *World*	3.4	2	1.4Gt/a
Monde h.pays soc. *WO.C.A.*	2.75	1.6	1.1Gt/a
Pays socialistes *C.P.E. areas*	0.65	0.4	0.3Gt/a

Fig.K. Ressources ultimes récupérables de gaz
Ultimate recoverable gas resources

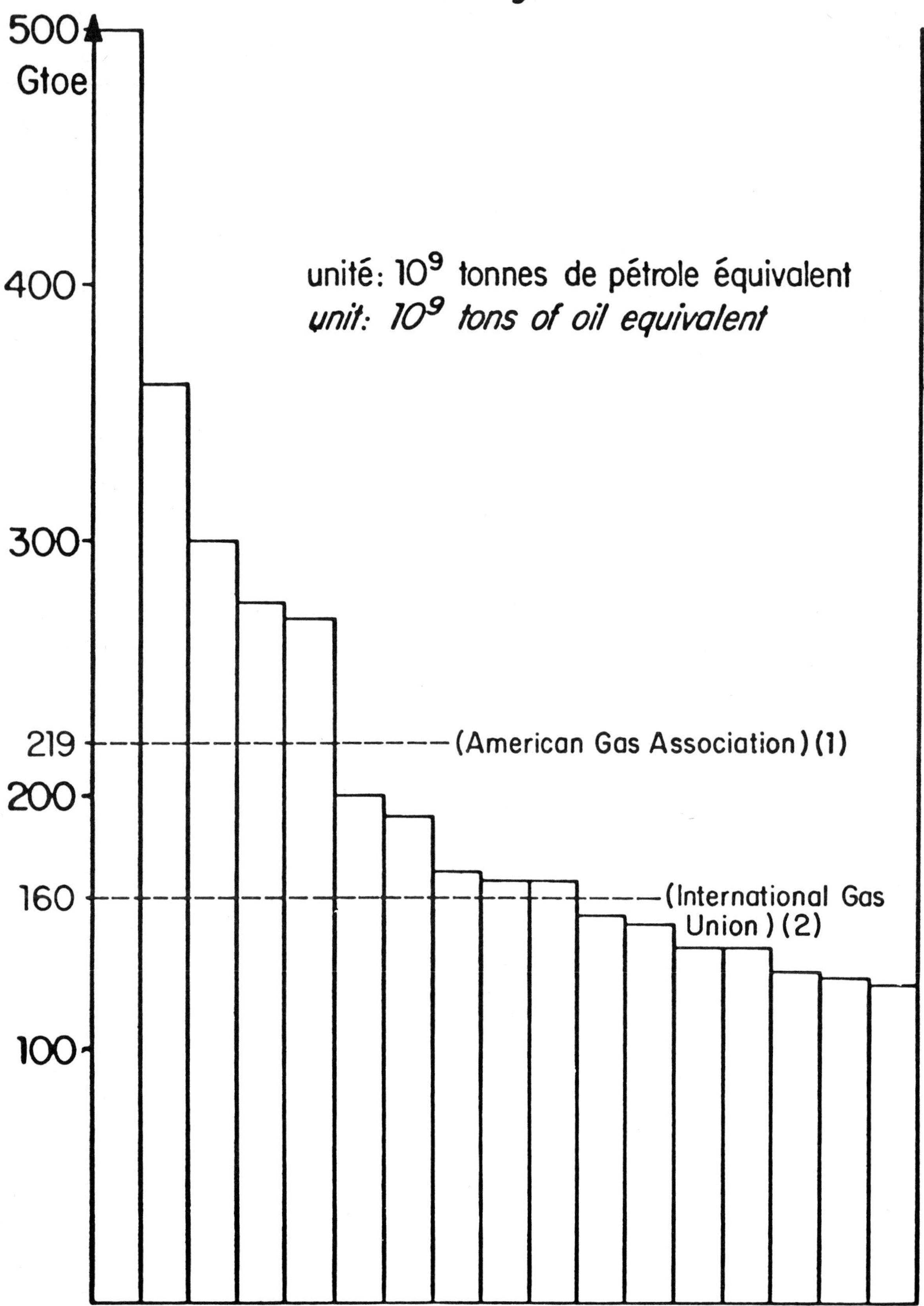

(1) Estimation de L.Fish
Estimated by L.Fish

(2) Estimation de L.J.Clark
Estimated by L.J.Clark

WORLD ENERGY CONFERENCE

PETROLEUM RESOURCES

QUESTIONNAIRE

Delphi-Type Poll of Companies and Experts on Ultimate Petroleum Resources

This Poll is aimed at obtaining the best possible range of approximation of ultimate recoverable petroleum resources. The first reply may subsequently be corrected by each evaluator in the light of the arithmetical averages of the answers received from others. Similar studies on all other primary energy resources are being made by other responsible groups of the World Energy Conference.

The results will eventually be incorporated in the final report of the Conservation Commission but before this report is written, the findings of each resource group will be subject to the critical analysis of experts at round-table meetings to be held during the 10th World Energy Conference in Istanbul in September 1977. These results will show, in the case of each question, the number of answers received, the average figures and the breakdown of the answers. The information will not be attributed to a source without the express agreement of the individual or company making the estimate.

DEFINITIONS

Ultimate Recoverable Petroleum Resources: means the resources still remaining to be produced (not including the 45 GT already produced) over the entire surface of the earth. This means the total of proven + probable + possible recoverable resources (including the Antarctic and deep offshore). It includes condensates and heavy oil fields (including the Orinoco Belt in Venezuela) but not reserves of tar sands and oil shales.

Since the concept of recoverable resources includes constraints of price and technology, for the present poll it will be assumed that the future price of petroleum will enable the best present and future technologies of exploration, production and recovery to be used and, consequently, that this price (technical cost i.e. amortization of exploration and production expenditures plus the direct production cost)*(Note i)* will tend toward $ 20 (1976) per barrel in the year 2000 (wellhead price) and that operators will receive the corresponding receipts.

Ideally, an answer to each question is desired, but if this is not possible, answers should be given to those questions where an estimate can be provided.

QUESTIONS

1. How much do you estimate the total ultimate petroleum resources of the earth to be?

1.1 Oil in place (recoverable cf. 1.2. plus unrecoverable resources)

- a) less than 750 GT *(Note ii)* ...*(Note iii)*
- b) between 750 and 1,000 GT ...*(Note iii)*
- c) more than 1,000 GT ...*(Note iii)*

1.2 Recoverable resources

- a) less than 300 GT *(Note ii)* ...*(Note iii)*
- b) between 300 and 500 GT ...*(Note iii)*
- c) more than 500 GT ...*(Note iii)*

NOTES:

(i) *for the sake of simplification, no consideration is given to net profit because its magnitude would not appreciably change the answers.*

(ii) *1 GT = 10^9 T and 1 T = 7.5 barrels of oil*

(iii) *Mark with an 'X' the category chosen. If available, please give a more precise estimate or range (high/low) of estimates.*

2. How do you break down recoverable resources (1.2.) among the different petroleum zones? *(Note i)*

			Percentage	Amount *(Note i)*
2.1	U.S.S.R. and Eastern Europe	*(Note ii & iii)*	%	
2.2	China	*(Note ii & iii)*	%	
2.3	United States and Canada	*(Note ii)*	%	
2.4	Middle East and North Africa	*(Note ii)*	%	
2.5	Africa (South of Sahara)	*(Note ii)*	%	
2.6	Europe	*(Note ii)*	%	
2.7	Latin America	*(Note ii)*	%	
2.8	Japan, Australia, New Zealand	*(Note ii & iv)*	%	
2.9	East and South Asia (including Pakistan, India and Vietnam)	*(Note ii & iv)*	%	
2.10	Deep offshore	*(Note v)*	%	
2.11	Arctic and Antarctic	*(Note v)*	%	
			100 %	

NOTES:

(i) *Even if you are unable to supply estimates for all regions, please give separate estimates for those regions for which you can supply information.*

(ii) *Omit deep offshore (more than 300 meters) and eventually Arctic and Antarctic zones.*

(iii) *2.1. and 2.2. may be combined.*

(iv) *2.8. and 2.9. may be combined.*

(v) *2.10 and 2.11 may be combined.*

3. How do you break down recoverable resources (1.2.) between onshore and offshore?

3.1	Onshore	%
3.2	Offshore	%
		100 %

4. Out of the total 1.2. for worldwide recoverable resources, what percentages do you consider could be brought into production in relation to the following technical costs per barrel *(Note i)* (amortization of exploration and production investments plus technical production costs, before taxes)?

4.1	less than $5	*(Note ii)*	%
4.2	between $5 and $12	*(Note ii)*	%
4.3	more than $12	*(Note ii)*	%
			100 %

5. Of the total 1.2. for worldwide recoverable resources what percentage of the total will have been produced or will be under production?

5.1	before 2000	%
5.2	after 2000	%
		100 %

NOTES:

(i) *For the sake of simplification, no consideration is given to net profit because its magnitude would not appreciably change the answers.*

(ii) *1976 U.S. dollars*

6. The Future Rate of Petroleum Reserves Discovery:

6.1 The world rate of discovery has been in the past roughly in step with the rate of growth of consumption (nearly 6% per year 1960/1972). Do you think that this ratio can be maintained, assuming the price of $20(1976) per petroleum barrel in the year 2000? i.e.:

6.1.1 A gross increase of reserves of 6/8GT per year around 1985.

6.1.1.1 YES

6.1.1.2 NO

6.1.1.3 If not, what amount seems more probable?

6.1.1.4 Out of this gross increase, what will be the net discovery rate (discovery of really new fields, excluding the re-evaluation of previously discovered fields)?

6.1.2 A gross increase of 12/16 GT per year around 2000.

6.1.2.1 YES

6.1.2.2. NO

6.1.2.3. If not, what amount seems more probable?

6.1.2.4. Out of this gross increase, what will be the net discovery rate (discovery of really new fields, excluding the re-evaluation of previously discovered fields)?

6.2 Up to now, discovery has been nearly proportional to prospection effort (seismic survey Km., meters drilled ....), technological progress compensating for the increasing difficulty of discovery. Do you think that, towards 1985/1990:

6.2.1 The proportion will remain the same? ....

6.2.2. It will be necessary to increase prospection effort in order to obtain the same quantity of oil? ....

6.2.3 If yes:

6.2.3.1 Less than twice current prospection effort?

6.2.3.2 More than twice current prospection effort?

7. Principal methods utilized for making these estimates:
(Please indicate which of these methods or combination of methods was used.)

7.1 Extrapolation methods based upon historical data.

7.2 Volumetric yields methods.

7.3 Combined methods (geological and statistical models).

8. Any Further Comment

9. The Future Recovery Rate of Known Petroleum Reserves:

9.1 In the context of the present projected price estimate of $20(1976) per barrel of oil in the year 2000, what in your opinion will be the maximum working percentage for the recovery of known oil reserves in the year 2000?

9.2 In your opinion, what will be the average rate of recovery at that time?

	<30%	30-40%	40-50%	50-60%	>60%
9.2.1 World Average Rate	:	:	:	:	
9.2.2 Average rate for the USA/Canada/Western Europe	:	:	:	:	
9.2.3 Average rate for the USSR/Eastern Europe	:	:	:	:	
9.2.4 Average rate for the rest of the World	:	:	:	:	

10. In view of completing the reply to question 6.2, concerning the cost of discovery, what do you think will be the development cost of one barrel/day capacity in constant 1976 dollars in the 1985-1990s?

	$000's b/d				
	<1	1 - 5	5 -10	10-20	>20
10.1 On land	:	:	:	:	
10.2 Conventional Offshore Zones (200m water depth)	:	:	:	:	
10.3 In Waters of Great Depth, the Arctic, Perma Frost Zones....	:	:	:	:	

11. Using an approximate calorific value GN 1000m³ = 1 Toe (or indicate the values used by you), what in your opinion is the percentage of recoverable ultimate natural gas world reserves in relationship to the ultimate recoverable world oil reserves?

$$\frac{\text{Ultimate recoverable gas reserves}}{\text{Ultimate recoverable oil reserves}} = \_\_\_\_\_\_ \%$$

World Energy Conference
Conférence mondiale dè l'énergie

34 ST. JAMES'S STREET,
LONDON SW1A 1HD

TELEPHONE 01-930 3966   TELEGRAMS WENERCON LONDON S.W.1
TELEX 21120 Mono. Ref. 1172

Our Ref: J.1/11/5 - 4292   3rd February, 1977.

Your Ref:

I am enclosing an analysis produced by the Institut Français du Pétrole of the replies which had been received by the end of December 1976, from 23 countries to whom the questionnaire had been addressed on the subject of ultimate world oil reserves.

Of the completed questionnaires received, sixteen came from the United States, (eleven companies, four experts, and one government agency) and seven replies originated in Europe (five companies and two independent experts).

We are awaiting replies from ten further companies to whom the original questionnaire was sent.

You will note that the replies received by the Institut Français du Pétrole are presented in such a way as to guarantee their anonymity, although comments from the replies are quoted without reference to their origin. The directors of the enquiry have tried to keep their own comments to a minimum.

The replies were classified and analysed so that you may be able to evaluate your own replies in relation to replies received from other organisations. This has been done with the object of giving you the opportunity to modify your initial estimates in the light of other replies received. The aim of the survey, as we stated initially, is to arrive, if possible, at a consensus of opinion, or at least at a number of areas of agreements.

cont........

Enclosed with this letter are:-

- A further copy of the original questionnaire entitled "Questionnaire No.2" with appended a redrafted question 6.1; this has been rewritten because some of the answers received to this question initially indicate that it could be open to different interpretations.

- three additional questions, numbered 9, 10 and 11, which, to judge from the replies already received, will be of interest. As in the original questionnaire, we request that you include a quantitative estimate. We should very much value your views.

- a note on heavy oil reserves.

To enable us to analyse this second and last series of replies in good time to prepare a preliminary report, we would appreciate receiving these before the 1st April 1977.

This preliminary report, after further criticism and scrutiny by other members of the Commission, will be sent to you by the end of July 1977 and will form the basis of the round-table discussion on petroleum resources at the Tenth World Energy Conference in Istanbul on the 21st September, 1977.

We would like to thank you again for your co-operation. Great interest has been shown by those to whom the questionnaire has been sent and their final compilation will represent the first world-wide survey by such a large number of experts on this vital subject.

Yours sincerely,

E.RUTTLEY.
Secretary-General.

Enc.

ER/ee.

NOTE - HEAVY OIL RESERVES

Some experts and companies encountered certain difficulties as concerns heavy oil reserves since the split between heavy oil and tar sands can be a matter of discussion. The answers to all questions will be considered as pertaining to conventional oil unless otherwise specified. In other words heavy oil of the type presently under production and classified as movable oil, for example, California are included but heavy oil resources not yet producible, for example, Orinoco Belt, Colombia, etc. which are sometimes considered as akin to tar sands are not included.

APPENDIX IV.

A. Comments on part 3. Unconventional Oil and 3.1 Oil Shales by Mr. W.A. ROBERTS, Executive Vice President, PHILLIPS PETROLEUM COMPANY, BARTLESVILLE, OKLAHOMA.

Part 3. Unconventional Oil

Extensive evaluation of these resources has not been made because they lack current economic importance. They are not sufficiently profitable under the existing economic environment to compete with conventional oil, and exploitation technologies either do not exist or have not yet been demonstrated adequately by commercial installations. Data on resource availability are as yet insufficient in many countries for their appraisal as potential energy resources. It is clear, however, that the total amount of hydrocarbon in these materials is extremely large and represents a major portion of the world's hydrocarbon reserves.

3.1 Oil Shales. Resources in place are currently estimated at 400 GT, of which about 5 to 10% could be considered for immediate exploitation. Although oil shale has been exploited in the United Kingdom, the Soviet Union, the People's Republic of China, Brazil and several other countries, it has been on a very modest scale. Perhaps the best chance of future commercial production is offered by the extensive oil shale deposits in western United States. Constraints to development include technology, environmental obstacles and governmental policies impacting economics. Superimpose these on high investments and uncertain economic situations and a slow and cautious development scenario is projected. Investment costs are about $20,000 per bbl/day with required selling price of refined shale oil lying between $20 and $25 (1976) per barrel to yield an acceptable rate of return. Crude shale oil could be produced $4 to $5 cheaper. Development work in modified in situ combustion offers the possibility of somewhat lower cost and might have an advantage in reduced ecological/environmental impact. Unless a constructive governmental program to assist the development of an oil shale industry is instituted, oil shale will probably not play any significant role in U. S. energy supply between now and the end of the century.

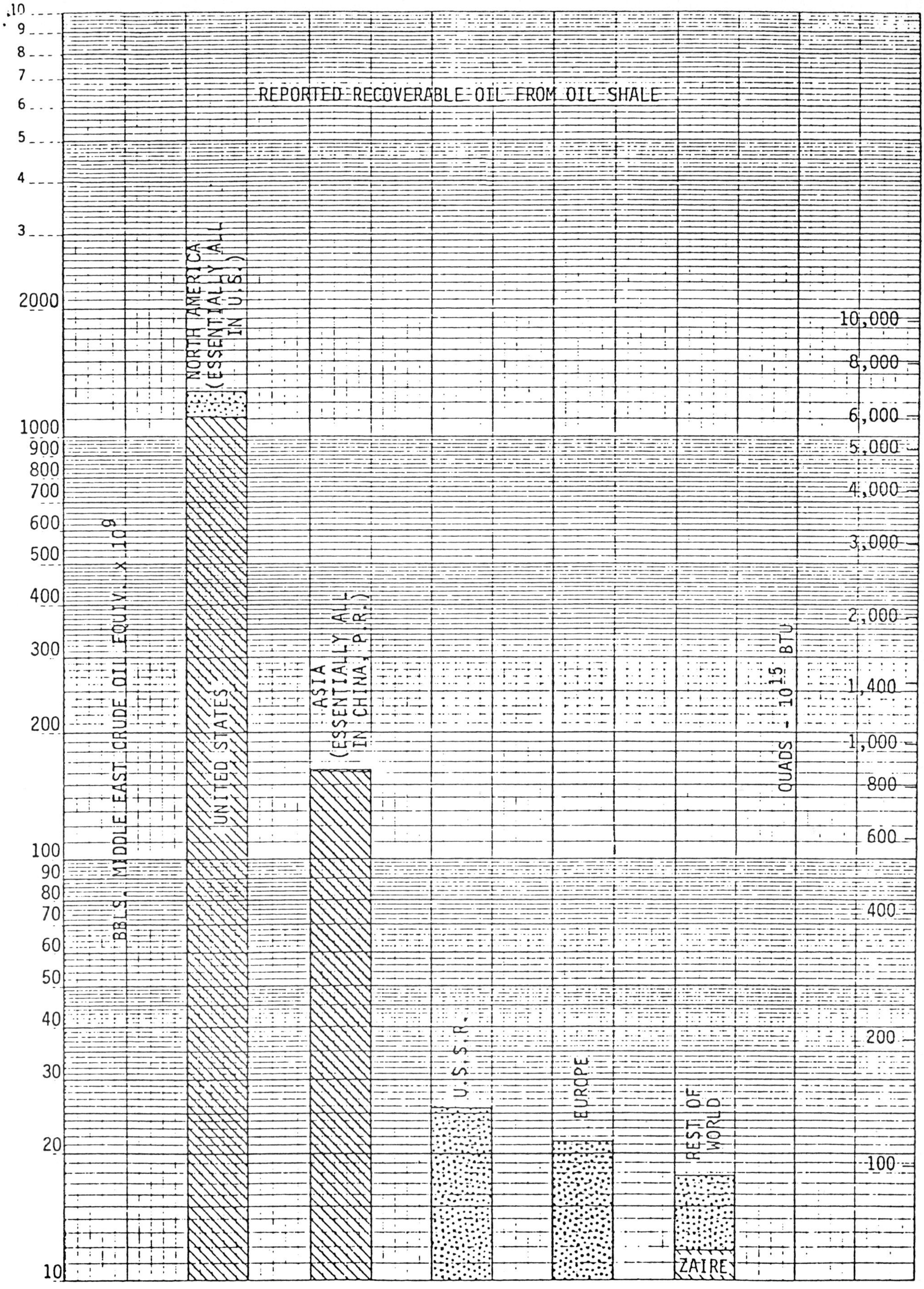

REPORTED RECOVERABLE OIL FROM OIL SHALE

BBLS. MIDDLE EAST CRUDE OIL EQUIV. x 10^9

QUADS - 10^{15} BTU

10, 9, 8, 7, 6, 5, 4, 3, 2000, 1000, 900, 800, 700, 600, 500, 400, 300, 200, 100, 90, 80, 70, 60, 50, 40, 30, 20, 10

10,000, 8,000, 6,000, 5,000, 4,000, 3,000, 2,000, 1,400, 1,000, 800, 600, 400, 200, 100

NORTH AMERICA (ESSENTIALLY ALL IN U.S.)

UNITED STATES

ASIA (ESSENTIALLY ALL IN CHINA, P.R.)

U.S.S.R.

EUROPE

REST OF WORLD

ZAIRE

PROPOSED REDRAFT 3.3 ENHANCED RECOVERY

3.3. Enhanced recovery may make it possible to increase the current average recovery rate of 25 to 30% to as much as 45 or 50% of the oil in many fields. Two recent American reports (Lewin 1976, NPC 1976) based on detailed surveys of existing fields emphasize both the hopes but also the drawbacks, i.e. high investments, uncertain technical results that may vary from single to triple or even quadruple, high production costs (it is between $15 and 20 per barrel that the production increase is the greatest) long time lags (10 to 12 years), and not very encouraging profitability with regard to the investment and technological risk.

Under auspices of the U. S. National Petroleum Council a Committee on Enhanced Oil Recovery conducted a study directed toward "An Analysis of the Potential for Enhanced Oil Recovery from Known Fields in the United States - 1976 to 2000." Results of this study for the U. S. should serve as a measure of what might be accomplished in fields throughout the world. The last two paragraphs are comments on results of the NPC 1976 report.

"The term "enhanced oil recovery" refers, in the broadest sense, to any method used to recover more oil from a petroleum reservoir than would be obtained by primary recovery. In primary recovery, naturally occurring forces, such as those associated with gas and liquid expansion or influx of water from aquifers, are utilized to produce the oil. Conventional secondary recovery methods, such as waterfloods, are considered to be "enhanced recovery" methods under this broader definition. Waterflooding of reservoirs, in which water is injected to supplement original reservoir forces and drive more oil to producing wells, currently accounts for about half of U. S. oil production. For this purpose, however, "enhanced oil recovery," or "EOR," is considered in a more narrow sense, and it is defined as: the additional recovery of oil from a petroleum reservoir over that which can be economically recovered by conventional primary and secondary methods."*

"The petroleum industry has conducted extensive research on enhanced oil recovery since the 1930's. As a result, several potential processes have been developed and field tested. Some of these processes are designed to recover the oil left in a reservoir after waterflooding or following other conventional secondary recovery processes. These EOR processes - usually the third type of recovery method employed in the reservoir - have been called "tertiary" recovery methods. Because some of the enhanced recovery processes may be used as an alternative to waterflooding or other conventional secondary recovery processes, the term "enhanced oil recovery" is considered to have a broader meaning than "tertiary" oil recovery."*

"The potential of enhanced oil recovery is of significant interest because of the number of fields in the U. S. and throughout the world to which it might be applied. While recovery in individual reservoirs is highly variable, the average recovery from conventional primary and secondary recovery methods in all U. S. reservoirs is expected to be only about one-third of the original oil in place, leaving nearly 300 billion barrels in currently known reservoirs. A portion of this remaining oil will constitute a target for enhanced oil recovery. The rest exists in unfavorable geologic or geographic regions or is so diffusely spread out in the reservoir rock that it very likely will not be recoverable by any process."*

"Three general classifications of EOR have shown significant promise. These classifications are: (1) chemical flooding; (2) carbon dioxide micible flooding; and (3) thermal methods. Of these, only one of the thermal methods (steamflooding) has been proven by several large-scale commercial applications. The other processes are receiving limited field testing at this time, but most of them have not been put into large-scale, commercial use because of their high cost."*

"The aggregate results for all U. S. fields for incremental ultimate recovery and potential producing rates are shown in Figures 1 and 2, respectively. A minimum 10 per cent Discounted Cash Flow Rate of Return (DCFROR) requirement has been used in these figures."*

"Incremental ultimate recovery from EOR processes increases with oil price, from less than 3 billion barrels at $5 per barrel to about 24 billion barrels at $25 per barrel, in constant 1976 dollars. At $5 per barrel, all EOR production is from thermal methods. At $10 per barrel, the contribution of carbon dioxide miscible flooding is about equal to that of the thermal methods. Chemical flooding has a low potential at $10 per barrel, but increases substantially with price, accounting for 9 billion of the 24 billion barrel total EOR at $25 per barrel."*

"The potential producing rate is also sensitive to oil price. The potential rate in 1985 varies from 0.3 million barrels per day at $5 per barrel to about 1.7 million barrels per day at $25 per barrel. Peak production (for most oil prices) from application of EOR processes to currently known reservoirs is projected in 1995 and ranges from about 0.25 million barrels per day at $5 per barrel to 3.5 million barrels per day at $25 per barrel. The uncertainty in potential producing rate at any point in time is larger than the uncertainty in ultimate recovery."*

Three points should be kept in mind with respect to results shown on Figures 1 and 2. (1) constant 1976 dollars, (2) 10% DCFROR, and (3) present free market price of oil is about $15 per barrel. (1) If oil prices rise faster than costs, results are understated. (2) If greater than 10% DCFROR is required, results are overstated. (3) Because the present free market price for oil is about $15 per barrel, results shown for the $15 per barrel price are probably the most meaningful.

Results for the $15 per barrel oil price show an ultimate incremental recovery of about 15 billion barrels, or about 3% of oil originally in place in known U. S. fields. This means that in the U. S.

on the average we can expect to increase ultimate recovery from about 32% to about 35% based on original oil in place in all fields. It should be recognized that EOR processes will not be applicable to all fields and where applicable recovery will be substantially better than 3% probably on the order of 10%. Further the potential additional oil from EOR processes in the U. S. and the world is significant although not as great as desired.

*From NPC 1976 Report.

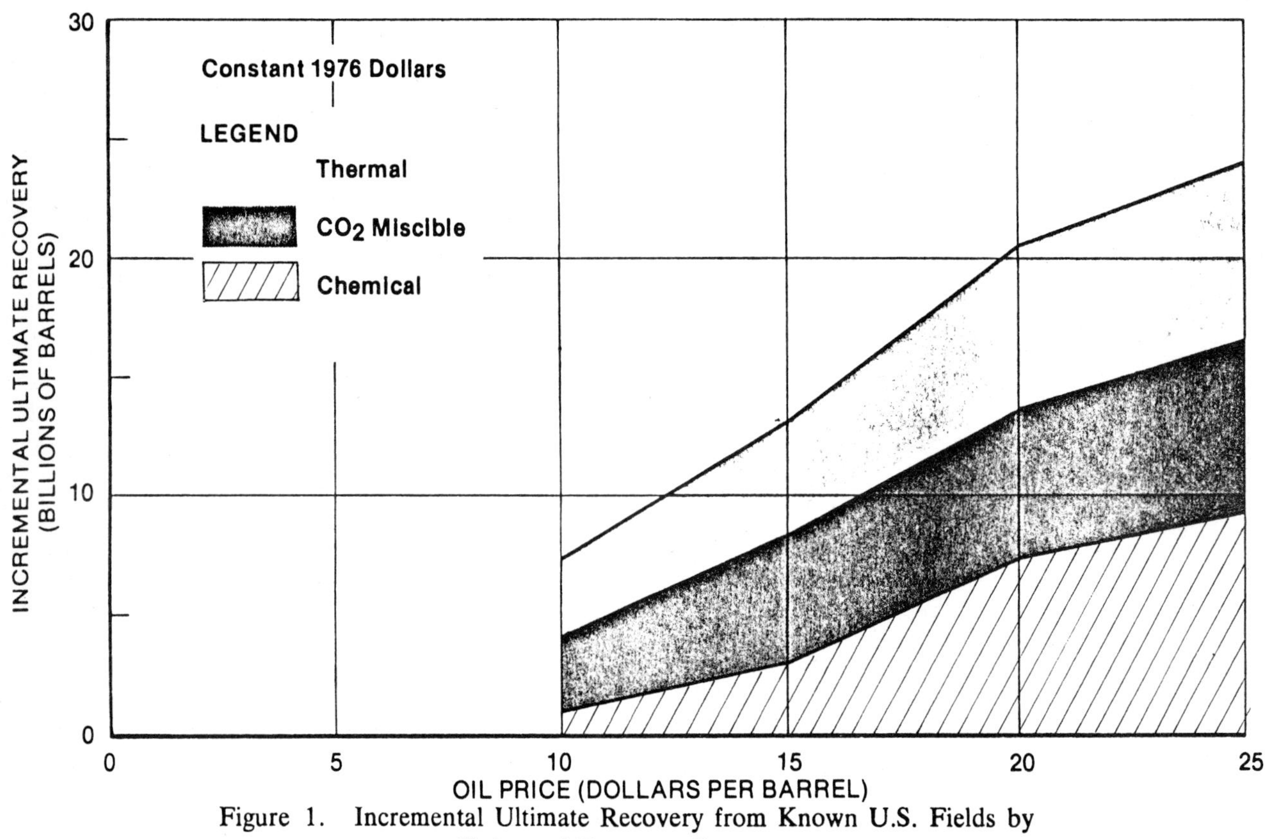

Figure 1. Incremental Ultimate Recovery from Known U.S. Fields by Enhanced Recovery Process.

*Recovery for Thermal at $5. Recovery for Chemical and CO_2 miscible flooding are zero at $5 per barrel and as shown at $10 per barrel. Recovery for intermediate prices between $5 and $10 per barrel has not been determined.

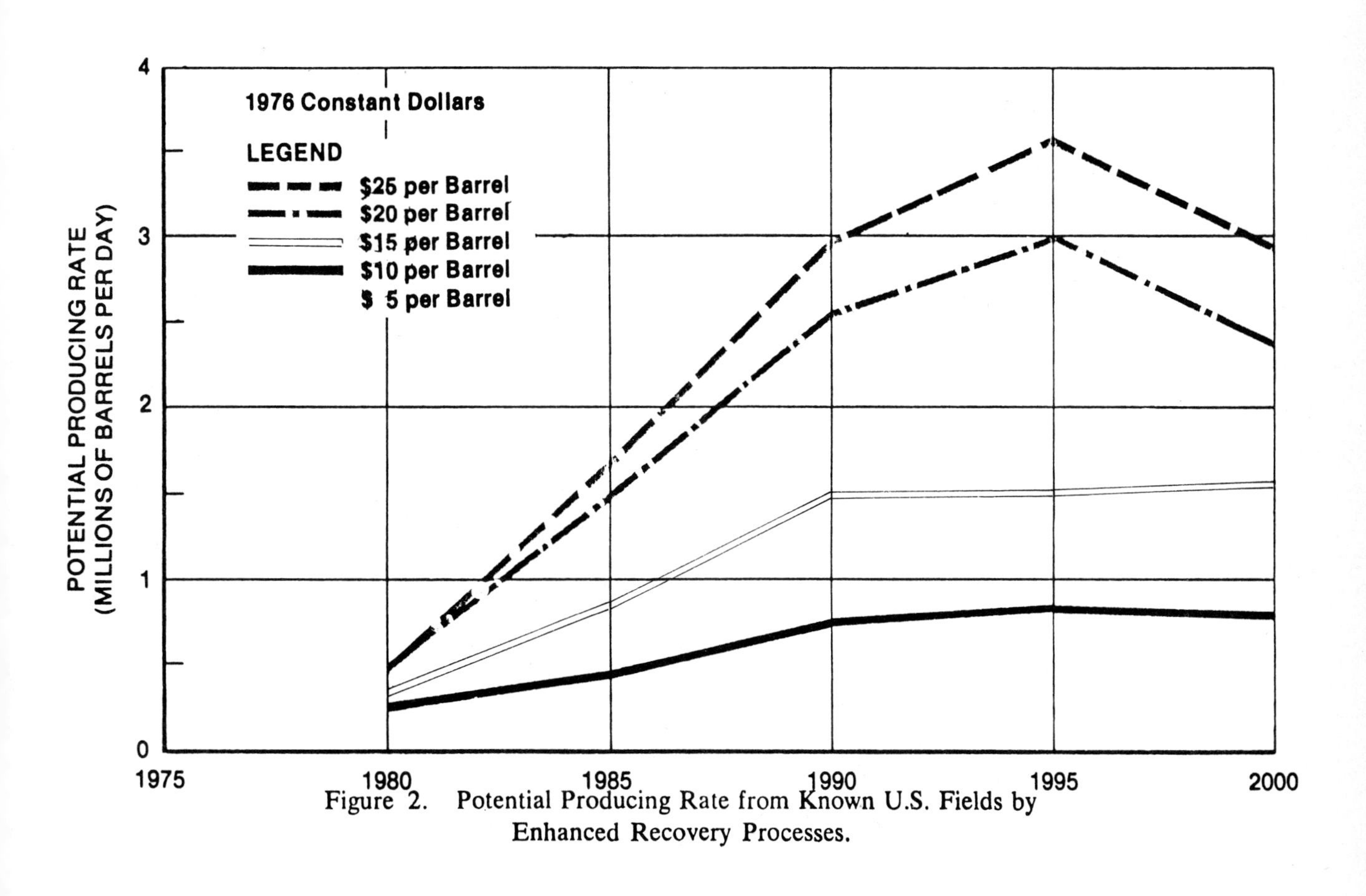

Figure 2. Potential Producing Rate from Known U.S. Fields by Enhanced Recovery Processes.

B. EXECUTIVE SUMMARY OF

"TAR SANDS AND HEAVY OILS" SECTION 3.2 OF 'WORLDWIDE PETROLEUM SUPPLY LIMITS'

W. N. SANDE
SYNCRUDE CANADA LTD.

The hydrocarbon associated with both tar sand and heavy oil deposits is high in specific gravity, high in sulphur and metals content. Geologists feel that this hydrocarbon has lost its light fractions as a result of previous contact with the atmosphere. The hydrocarbon occurs in soft sandstone and sandy rocks. Hydrocarbon that does not flow at ambient temperatures is called bitumen, whereas hydrocarbon that does flow at ambient temperatures is called heavy oil.

These oils exist in abundance on the earth with a wide geographic distribution. Reserves are several-fold those of conventional oil. Ninety percent of the world-wide tar sand/heavy oil resources can be found in three countries - Canada, Venezuela and Russia.

Approximately 300 billion tonnes (2100 billion barrels) divided up among four giant fields: Orinoco (Venezuela) with 700 billion barrels; Athabasca (Canada) 600; Olenek (USSR) 600; Cold Lake (Canada) 160.

Approximately 27 billion tonnes (190 billion barrels) between eight large fields: two in Canada (Wabasca and Peace River) with 160 billion barrels; five in the United States with 27 billion barrels, and one in Madagascar (Bemoulang) with 1.7 billion barrels.

Approximately 120 million tonnes (860 million barrels) among nine medium-sized fields, including four in the United States with 320 million barrels; one in Albania (Selenizza) with 370 million barrels; one in Venezuela with 62; one in Trinidad with 60; one in Rumania with 25, and one in the Soviet Union with 24.

The technology for recovering bitumen/heavy oils falls into two general categories: (a) Surface Mining, and (b) In-Situ Recovery.

Only five to ten percent of the reserves are potentially recoverable by surface mining methods. The remainder would be recovered by in-situ techniques, estimated recovery 30 to 50 percent of reserves.

The hydrocarbon recovered by either method must be further upgraded to a synthetic crude oil by either removing carbon (coking) or adding hydrogen (hydrocracking). The upgrading process desulphurizes and demetallizes the bitumen/ heavy oil and enhances its transportation characteristics.

Both recovery methods are capital and labour intensive to construct and operate. In Canada, the surface mining investment costs $22,000 (1976) per barrel per day. Direct operating costs are in the range of $5-6 (1976) per barrel of synthetic crude. Similar investment and operating costs would be incurred for in-situ recovery processes. The surface mining technology requires considerable investment and operating cost to cope with the major materials handling problem, whereas in-situ technology eliminates the materials handling problem and substitutes the need to input energy into the formation to improve the reservoir transport of the hydrocarbon. Both processes will employ the same upgrading technology to produce similar synthetic crudes.

Commercially, two surface mining processes are operational or near-operational in Canada. Great Canadian Oil Sands (GCOS), in operation since 1967 with a production capacity of 50,000 barrels per day of synthetic crude and the Syncrude Canada Ltd. Project designed for 125,000 barrels per day of synthetic crude to commence operation in 1978.

In-situ technology is in the pilot stage in Canada, Venezuela and Russia. The volume of reserves potentially recoverable by this technology is the incentive for developing economic recovery methods. In addition, the potential to produce in excess of 125,000 barrels per day with a single facility adds to the incentive.

In Canada, over $250 million is being spent on five different recovery processes by AOSTRA (Alberta Oil Sands Technology and Research Authority) and Industry. An additional $100 million has already been spent by Industry in an attempt to commercialize a process. 1985 is the earliest a commercial facility of 125,000 barrels per day could be operational.

Venezuela's pilot program has been reported to be lagging that of Canada's. They target commercial production (12.5 MT/year or 250,000 barrels per day) to be operational by 1990.

The Russian technical effort is every bit as active as that of Canada and Venezuela. They operate a combined underground mining operation and thermal extraction process at their Yarega field.

* * * * *

THE FUTURE FOR WORLD NATURAL GAS SUPPLY

W. T. McCormick, Jr, L. W. Fish, R. B. Kalisch and T. J. Wander
American Gas Association

INTRODUCTION

This is a summary of a report prepared for the World Energy Conference as part of an overall examination of future world supply and demand for energy which presents estimates of the world conventional natural gas production capability by regions through the year 2020. Production capability, as opposed to future production, is estimated since it is not possible to predict the economic, political, and other factors which might influence future production decisions.

The full report contains three sections:

- Results and conclusions including estimates of world gas production capability for 1985, 2000, and 2020.
- A discussion of data sources including proved gas reserves and undiscovered remaining gas resources, a description of the methodology for estimating production capability, and a review of key assumptions used in the study.
- Graphic presentations of gas production capability estimates for each region of the world accompanied by a short discussion of significant issues of interest.

RESULTS AND CONCLUSIONS

- There is a worldwide capability for substantially increasing production of conventional natural gas in the next ten years and sustaining production levels well above today's level at least until the year 2020.
    - While annual world conventional natural gas production is now only about 50 Exajoules (EJ), proved reserves are estimated at about 2,500 EJ and remaining undiscovered resources are estimated at about 8,100 EJ.
    - Cumulative worldwide conventional production of natural gas to 1975 is estimated at about 929 EJ or about 40% of presently estimated proved reserves and only 11% of remaining undiscovered gas resources.
- Even at an annual world natural gas production rate of double the present rate (i.e., 100 EJ/year vs. the present 50 EJ/year), the estimated world remaining conventional natural gas resource base would be sufficient to sustain production at or near this level for at least another fifty years.

- This conclusion does not assume any production from the numerous sources of unconventional and supplemental sources of natural gas from geopressured resources, from tight gas formations, from coal beds, from shales, and from biomass. They represent an additional substantial gas resource base estimated in the range of several thousand to tens of thousands of Exajoules.

- Under a gas pricing scenario which would allow a natural gas price of $20/bbl crude oil equivalent (1974 dollars) in the period after 1985, it is estimated that world gas production could rise to about 77 EJ by 1985 and to about 143 EJ by the year 2000.

  - At these production rate increases (4.4% per year through 2000), it is estimated that production would peak shortly after the year 2000 and decline to about 125 EJ by 2020. By that time, about 50% of the presently estimated remaining gas resource base would have been produced.

  - Key areas of the world where substantial potential exists for greatly increasing production over the next decade include the OPEC groups and the U.S.S.R.

  - Production capability would peak in only two regions of the world prior to the year 2000, i.e. in the more mature, developed areas of North America and Western Europe. By contrast, several areas of the world, notably OPEC Group 2, could still be increasing gas production by 2020.

- Under a medium gas pricing scenario, which would only allow $14/bbl crude oil equivalent (1974 dollars), production capability in the period from now until 2020 is estimated to be lower than in the $20/bbl scenario, although only slightly lower until after the year 2000. The small difference in production capability between the two scenarios until after the year 2000 is because prices in the range of $20/bbl are not yet expected to be required in order to produce the more easily recoverable resources.

- Finally, the gas production capability estimates shown in Table 1 do not include potential production from non-conventional sources sources such as gas from coal beds. sured resources, shale, biomass, etc., which could add significantly to world gas production in the post-2000 years.

Table 1

Estimated Natural Gas Production Capability by Region

(Exajoules)

Region	1976* Actual	1985 High	2000 High	2000 Medium	2020 High	2020 Medium
North America	23.0	29.7	27.3	26.6	10.7	7.5
Western Europe	6.4	9.6	8.7	8.4	2.2	1.6
J A N Z	.3	.4	2.1	2.1	4.6	4.5
USSR; East Europe	12.8	21.8	55.7	55.6	28.5	25.3
China; Other Asia	1.4	1.7	2.9	2.9	6.1	6.0
OPEC, Group 1	.5	7.0	18.1	18.1	17.7	16.4
OPEC, Group 2	3.4	4.9	21.3	21.3	45.6	44.6
Central America	.9	1.1	2.3	2.2	1.6	1.4
South America	.8	1.1	2.2	2.2	4.8	4.7
Middle East	.1	.5	1.0	1.0	.3	.2
North Africa	.2	.3	.5	.5	.5	.4
Africa, South of the Sahara	.1	.1	.2	.2	.1	.1
East Asia	.1	.1	.2	.2	1.6	1.6
South Asia	.3	.5	1.0	1.0	.7	.5
WORLD TOTAL	50.3	76.8	143.5	142.3	125.0	114.8

*Rounded. Based on Oil & Gas Journal, Vol. 75, No. 4 (Jan.24, 1977), pg. 95.

ESTIMATES OF PROVED RESERVES AND UNDISCOVERED RESOURCES

Estimates of proved reserves, remaining undiscovered resources, and cumlative gas production developed in this study are shown in Table 2. Proved reserves were aggregated by region from individual national estimates published by World Oil magazine.[1] Aggregate estimates of undiscovered gas resources were based primarily on figures published by the Institute of Gas Technology.[2] Allocation by region was based on disaggregated data available from other sources and, when not available, on the ratios of proved reserves, presuming that gas resources are generally larger in areas where gas has been found. However, allowances were made for the considerable regional differences in the degree of exploration and development.

A comparison of these estimates of proved reserves and undiscovered resources with other sources show an average range of 7 percent for proved reserves and generally about 25 percent for undiscovered resources, indicating the high degree of uncertainty concerning resources in unexplored or undeveloped regions.

Cumulative production estimates by region were based primarily upon U.N. data and individual national data. However, production data generally exclude flared gas and, although efforts were made to estimate these volumes, the cumulative production estimates can be realistically regarded as good approximations. However, the total of 929 is in close agreement with an estimate by the Institute of Gas Technology of 903.5 to 930.7 EJ.

PRODUCTION CAPABILITY MODEL

The production capability estimates in this study, summarized in Table 1, are estimates of the potential production a region could attain if exploration and development of gas resources proceed at a historical pace similar to U.S. experience. In developing the production capability estimates, no consideration was given to whether the potential production would be consumed within the producing or other regions. The estimates should only be interpreted to mean that there is the potential to achieve a certain level of production by a specific year depending on the assumed price scenario developed

---

[1] World Oil, Vol. 183, No.3(August 15, 1976), p.44.

[2] Parent, Joseph D. and Linden, Henry R. "A Survey of United States and Total World Production Proved Reserves and Remaining Recoverable Resources of Fossil Fuels and Uranium as of December 21, 1975," Chicago: Institute of Gas Technology,1977.

Table 2

World Gas Reserves, Resources, and Cumulative Production Estimates* - 1975

(Exajoules)

Region	Proved Reserves	Undiscovered Resources	Cumulative Production
North America	310	1640	637
Western Europe	152	315	43
JANZ	41	232	2
U.S.S.R.; East Europe	795	2222	140
China; Other Asia	21	380	2
OPEC, Group 1	250	1042	50
OPEC, Group 2	687	1675	34
Central America	20	127	10
South America	22	277	5
Middle East	15	30	1
North Africa	8	32	1
Africa, South of the Sahara	3	12	1
East Asia	21	120	1
South Asia	16	43	2
TOTAL	2362	8147	929

*May not sum to total due to rounding.

by the Conservation Commission of the WEC for this study. This potential assumes an active exploration and development program, and the corresponding evolution of a domestic or export market for the gas.

Near-term production capability estimates (i.e., to 1985) are based partly on extrapolations of historical production records, announced development plans, and projected production figures, when available.

Farther term production capability estimates for each region are based on a generalized model developed from U.S. data. The model, discussed at length in the full report, calculates production based on a generalized model developed from U.S. data. The model, discussed at length in the full report,calculates production based on a proved reserves to production ratio (R/p) and a finding rate (% of undiscovered resources transferred into proved reserves). These parameters have been found to vary in a predictable manner as functions of the percentage of the resource base cumulatively produced.

Under the high price scenario (i.e., $20/bbl of crude oil in 1974 dollars), it was assumed that the total remaining recoverable resource base could eventually be produced. This price is about double the current unregulated prices in the U.S., a relatively more mature region of the world from a cumulative production standpoint.

Under the medium price scenario (i.e., $14/bbl of crude oil in 1974 dollars), the assumption was made that only 75% of the orginal recoverable resource base could be recovered. This assumption was based on the results of a study of recent U.S. drilling activity which showed the less costly drilling ocurring in areas estimated to have contained about 75% of the original gas in place.

From the generalized model individual regional production capability curves were developed as summarized in Table 1. By 2020, the sensitivity of production to price is evident in the more mature regions. For example, in North America the production differential under the two price scenarios is 30%. As resource development matures the medium price case therefore results in significantly lower production with a resultant increase in the demand of alternative fuels such as unconventional or supplemental gas supplies, oil or electricity from coal or nuclear sources. Currently, in the U.S. end-use electricity rates are three times greater than the $20/bbl high price case indicating that production at the high price case will still remain economically advantageous.

UNCONVENTIONAL GAS RESOURCES

The gas resource and production estimates provided do not include unconventional and supplemental sources of gas, e.g., gas produced from coal, biomass, and geopressured resources.

Knowledge of the extent of the resource base and the potential for recovering these resources on a global scale is limited. However, as shown in Table 3, the estimates for just the U.S. alone show substantial volumes of gas. New technological advancement which would allow economic recovery of these resources could add significantly to world production capability.

Supplemental Sources of Natural Gas -- U.S.

Source	Estimated Volume in Place (EJ)
Coal-bed Degasification	325-870
Devonian Shale	545-650
Tight Formations	650
Geopressured Gas	3,200-54,400

Finally, technological improvements in enhanced recovery techniques can be expected which will boost upwards the gas resource recovery factor from its present average of about 70-80% thereby expanding the recoverable volumes of conventional gas resources.

AN APPRAISAL OF WORLD COAL RESOURCES AND THEIR FUTURE AVAILABILITY

W. Peters and H.-D. Schilling
with the assistance of W. Pickhardt, D. Wiegand and R. Hildebrandt
Bergbau-Forschung GmbH, Central Research and Development Institute of the Hard Coal Mining Industry of the Federal Republic of Germany

Remark:

The Unit applied is 1 ton coal equivalent (1 t. c. e.). This corresponds (in average) to the calorific value of 1 metric ton of hard coal. Its definition is the following:

1 t. c. e.	=	7	Gigacalories
	=	29. 30	Gigajoule
	=	27. 78	MMBTU

Contents

0. Summary of Findings and Conclusions

1. The Basic Situation

2. Aims and Objectives

3. Execution of the Study

4. Estimation of the World's Coal Resources and Reserves

4.1. Estimation Criteria

4.2. Estimating the Coal Resources and Reserves

4.3. Evaluation of the Data

4.3.1. Comparison with Earlier Data of the WEC
4.3.2. Evaluation of Figures for Resources and Reserves
4.3.3. The Geographical Distribution of the Coal Resources and Reserves

5. The Availability of the Coal Reserves

5.1. Preliminary Remarks and Conduct of the Investigation

5.2. Estimated Possible Development of Coal Production in the Main Coal-Producing Countries

5.2.1. Estimated Future Coal Production
5.2.2. Evaluation of the Production Data
5.2.3. Possible Bottlenecks in the Sufficient Availability of Coal
5.2.4. Global Availability (Exportavailability)

6. Economic-Statistical Estimation of the Possible Trend in Production, based on Different Reserves Data until 2020

7. Possibilities of an Increased Coal Output

0. Summary of Findings and Conclusions

According to the mandate of the Conservation Commission, it was the aim of this study to assess the contribution which coal could make to future energy supplies. In this assessment, attention is concentrated on the appraisal of global coal resources and their future availability. Also possible constraints hindering an increase in coal production had to be considered. The study's findings, and its conclusions are summarized as follows:

1. Resources and Reserves

1.1. Present-day geological world coal resources are estimated at more than 10 000 x 10^9 tons of coal equivalent (t. c. e.).

1.2. Those coal reserves currently estimated as technically and economically recoverable amount to c. 640 x 10^9 t. c. e.

2. Production

2.1. According to the present planning of the coal producing countries, and, also, additional estimates based primarily on present coal reserves, in the year 2020 a world coal production of 8.8 x 10^9 t. c. e. could be achieved, if the necessary actions are taken in time.

2.2. This level of production would mean more than tripling today's world coal production (2.6 x 10^9 t. c. e.). It would require an average annual growth rate of 2. 7 % during the period 1975-2020 (as compared with 2. 6 % during the period 1860-1975 and 2. 2 % during the period 1950-1975).

2.3. Provided, a sufficient economical incentive will be given for the step up of coal production and especially for an increase of coal exports, some experts consider a total annual coal output of about 13 000 million tons coal equivalents (t. c. e.) possible for the year 2020. This would mean at the same time an increase of the technically and economically recoverable reserves mentioned in section 1. 2.

3. Possible constraints on the increase of coal production

In order to achieve this production, a number of constraints must be overcome, which might hinder a substantial increase in coal production in many parts of the world.

These major obstacles are:

3.1. The recruitment of a sufficient number of qualified miners and engineers.

3.2. The construction of a suitable infrastructure and of adequate transportation facilities.

3.3. Various environmental problems, which need to be solved, both in production and in consumption.

3.4. The fact, that at present potential markets for coal are not yet sufficiently being developed in many parts of the world, since other sources of energy are still offered at lower prices. This means that there is also a lack of interest on the part of potential investors to commit themselves to the development of coal.

3.5. The long lead times required for opening up new mines, establishing the necessary infrastructure, transportation facilities etc.

4. Regional and global availability of coal

4.1. Most main coal producing countries plan a substantial increase of their coal production capacity, which is apparently capable of meeting their own national demands. A sufficient regional availability for coal in the main coal producing countries, therefore, may be expected, although in some cases the constraints mentioned above could lead to planned productions being reduced.

4.2. However, most of the countries, so far, seem to plan their future coal production mainly according to their own future requirements. Some countries, which might be capable of producing large quantities of coal, do not, as yet, seem willing to undertake the extensive coal mining and related development of transportation system required for an increase of exportation. According to present planning and current estimates of future coal exports, the average export quota of coal lies between c. 7 % and 10 %of the production figures.

Apparently, this rate is much too low to meet the rate of demand of coal importing countries, which is itself proportionately higher. Nor is it sufficient for developing an extensive international coal trade. Current efforts are therefore inadequate for securing a sufficient global availability of coal.

4.3. An increased economical incentive for a higher coal output provided, experts consider a maximum coal output of 13 000 million t.c.e. to be possible for the year 2020. About 5 000 million tons thereof could be available for exports provided there is suitable infrastructure. Such a volume would amount to nearly 40 % of the total coal output.

5. Policy conclusions

5.1. The world possesses abundant coal occurences. One may also assume, in addition to currently estimated resources and reserves, that there is a considerable "potential behind the potential".

5.2. Coal, therefore, could substantially contribute to future energy supplies. It could also reduce the risk of possible gaps between supply and demand in energy, created by diminishing production rates of oil and gas or by difficulties with nuclear energy.

5.3. It is true that a number of obstacles might hinder the timely and adequate use of the potential offered by coal. However, one should at the same time make it clear that these obstacles can be removed, provided that appropriate action is taken.

5.4. At present, the main problem seems to be that potential markets for coal are not yet sufficiently being developed, since other energy sources still are plentiful and offered at lower prices. This again leads to a lack of willingness on the part of potential investors to commit themselves to the development of coal.

5.5. Considering the long lead times required for an extensive production and use of coal (and this includes policy decisions, investment commitments, coal mine development, transportation etc.), one cannot merely rely on a future market which might be more favourable to coal.

5.6. Action must be taken now, if the maximum use of the potential offered by coal is to be made. Therefore, appropriate policy decisions by governments and by coal consumers are imperative. These decisions should be aimed to enable potential coal consumers to commit themselves to long-term contracts. This, again, would enable and encourage potential investors to commit themselves in time, and without unacceptable risks, to the necessary development of coal.

1. The Basic Situation

The studies carried out by various world institutions during the past few years on the future trends in the world's energy consumption and demand, make one fear that in approximately 10 to 15 years, a worldwide gap in the supplies of oil and natural gas will have to be reckoned with. The result of this will be an increased utilization of other fuels, the only substitutes that can be considered to replace these for the foreseeable future, being the vast resources of nuclear energy and coal.

The major part of the worldwide consumption of secondary energy fuels is based today on hydrocarbons. For this reason, the world energy industry of the future will accordingly also have to gear itself to the production of refined hydrocarbons. The reasons for this are to be found in the currently employed technologies, the market structure, and in efforts to maintain market-orientated supplies and complying with the wishes of the consumers. Whereas in the foreseeable future, however, nuclear energy can only be used for the generation of electricity, in coal we have a raw material at our disposal, whose utilization and refining potential is hardly attained by any other material. The product range that can be achieved by it extends from electricity via gaseous and liquid fuels and chemical feedstocks right through coke as to activated coal, which can be employed to advantage particulary in the field of environment protection.

While this is the advantage of coal, there is, however, also a disadvantage: the mining of coal requires varying degrees of technical skill. Furthermore, coal is a low-hydrogen, high-molecular solid substance that is intergrown with mineral substances, so that the refinement processes are only possible with the application of relatively sophisticated technology.

2. Aims and Objectives

The need for the increased use of coal raises the question, however, whether and to what extent, based on today's assessment of the situation, an adequate supply of coal is to be regarded as ensured in the medium and long term and with regards to both regional and global demand. Naturally, the assured supply of coal will depend, first and foremost, on the extent and volume of the existing resources. In the light of the above-mentioned aspects of the technological difficulties involved in coal production, and those of the problems inherent in the quality characteristics of coal, the term "availability of these resources" acquires an importance of equal degree. It is therefore the aim of this study not only to carry out an estimation of the world's resources and reserves, but also to make statements on their future availability.

3. Execution of the Study

The Study was divided into two focal points of emphasis: by means of an in-depth analysis of the pertinent literature, an attempt was made to obtain a broad basis for comparative statements on the world's coal resources and reserves and their availability. With the aid of a world-wide questionnaire compaign made at the same time, supplemented by personal talks with the main coal producing countries, an effort was furthermore made by means of specific questions to obtain up-to-date and authentic data that would be of particular importance for estimating the coal resources and reserves in terms of their availability. In the course of this work it was found that the greater part of the countries were not yet in a position to make statements on their future coal industry beyond the year 2000. Despite this, it was found possible to carry out estimates for the period up to the year 2020 (mainly based on data from literature). The results of these investigations already reveal significant trends in the future availability of coal.

4. Estimation of the World's Coal Resources and Reserves

4.1. Estimation Criteria

Within the framework of this Study the following differentiations are made:

1. Geological Resources
   These are understood to mean resources that may become of economic value to mankind, at some time in the future.

2. Technically and Economically Recoverable Reserves
   This term covers reserves that can be regarded as actually recoverable under the technical and economic conditions prevailing today.

Within this context the following maximal depth limits were specified:

Geological resources of

hard coal:	2 000 m
brown coal:	1 500 m

Technically and economically recoverable reserves of

hard coal:	1 500 m
brown coal:	600 m

The minimum seam thicknesses for the technically and economically recoverable coal reserves is generally in the range of around 0.6 m for hard coal, and around 2 m for brown coal. In a number of countries, however, there are considerable deviations from the above-mentioned figures; the figures for depth range are frequently substantially lower.

Differentiating in terms of rank between brown coal and hard coal is problematic, however, since in the natural progression of rank the transitions are fluid. For the purposes of this Study the coalification limit between hard coal and brown coal was fixed at 23. 76 MJ/kg (equivalent to 5 700 kcal/kg), related to air-dried, ash-free substance. By doing so, this takes account of the distinction made usually in several coal-producing countries between "bituminous coal and anthracite" and "subbituminous coal and lignite". Since it was not possible for many countries to break down their resources and reserves according to calorific values, the data given were converted by us, for reasons of facilitating comparability, into t. c. e. in accordance with the relevant conversion factors as used in the UN-statistics.

4. 2. Estimating the Coal Resources and Reserves

Table 1 gives a review of the coal resources and reserves, broken down according to continents and countries. Table 2 lists these same data but grouped only by continents. The total resources of hard coal and brown coal amount to 10, 125 x 10^9 t. c. e. ; the reserves currently regarded as being technically and economically recoverable come to 636 x 10^9 t. c. e. Thus, the reserves account for around 6. 3 % of the resources.

4. 3. Evaluation of the Data

4. 3. 1. Comparison with Earlier Data of the WEC

The data presented in this report differ from the values that were published by the World Energy Conference in its "Survey of Energy Resources" of 1974 and 1976. In Table 3 as well as in Fig. 1 attached these data are compared with one another.

It is be seen that both the geological resources and the reserves are following an upward trend, the jump in the resources between 1976 and the compilation of this report, namely approx. 10^9 t. c. e., being a relatively large one. However, the geological resources and the technically and economically recoverable reserves are increasing nearly in the same ratio. One of the reasons for the higher figures is apparently to be found in the fact that many countries, following the changes in the energy economy situation in the years 1973/74, have carried out a re-evaluation of their coal resources (especially of the reserves). Among others, countries like the United Kingdom, South Africa and Botswana, for example, have considerable increased their figures for the technically and economically recoverable reserves.

Continent: America

Country	Geological resources in 10^6 t.c.e.		Technically and economically recoverable reserves in 10^6 t.c.e.	
	h.c. +)	b.c. +)	h.c.	b.c.
Argentina	--	384	--	290
Brazil	4 040	6 042	2 510	5 588
Canada	96 225	19 127	8 708	673
Chile	2 438	2 147	36	126
Columbia	7 633	685	397	46
Mexico	5 448	--	875	--
Peru	1 072	50	105	--
USA	1 190 000	1 380 398	113 230	64 358
Venezuela	1 630	--	978	--
Other countries	55	5	--	--
Total	1 308 541	1 408 838	126 839	71 081

Continent: Europe

Country	Geological resources in 10^6 t.c.e.		Technically and economically recoverable reserves in 10^6 t.c.e.	
	h.c. +)	b.c. +)	h.c.	b.c.
Belgium	253	--	127	--
Bulgaria	34	2.599	24	2.179
Czechoslovakia	11 573	5 914	2 493	2 322
Fed. Rep. of Germany	230 300	16 500	23 919	10 500
France	2 325	42	427	11
German Democratic Rep.	200	9 200	100	7 560
Greece	--	895	--	400
Hungaria	714	2 839	225	725
Netherlands	2 900	--	1 430	--
Poland	121 000	4 500	20 000	1 000
Romania	590	1 287	50	363
Spain	1 786	512	322	215
United Kingdom	163 576	--	45 000	--
Yugoslavie	104	10.823	35	8 430
Other countries	309	130	58	57
Total	535 664	55 241	94 210	33 762

+) h.c. = hard coal (bituminous coal and anthracite)
b.c. = brown coal (subbituminous coal and lignite)

Continent: Africa

Country	Geological resources in 10^6 t.c.e.		Technically and economically recoverable reserves in 10^6 t.c.e.	
	h.c. +)	b.c. +)	h.c.	b.c.
Mozambique	400	--	80	--
Nigeria		180	---	90
Republic of Botswana	100 000	--	3 500	--
Republic of South Africa	57 566	--	26 903	--
Rhodesia	7 130	--	755	--
Swaziland	5 000	--	1 820	--
Zambia	228	--	5	--
Other countries	2 390	10	970	--
Total	172 714	190	34 033	90

Continent: Australia and the Pacific South Sea

Country	Geological resources in 10^6 t.c.e.		Technically and economically recoverable reserves in 10^6 t.c.e.	
	h.c. +)	b.c. +)	h.c.	b.c.
Australia	213 760	48 374	18 128	9 225
New Zealand	130	660	36	108
Other countries	--	--	--	--
Total	213 890	49 034	18 164	9 333

Continent: Asia

Country	Geological resources in 10^6 t.c.e.		Technically and economically recoverable reserves in 10^6 t.c.e.	
	h.c. +)	b.c. +)	h.c.	b.c.
Bangladesh	1 649		517	2
China (PR)	1 424 680	13 365	98 883	n.a.
India	55 575	1 224	33 345	355
Indonesia	573	3 150	80	1 350
Iran	385	--	193	--
Japan	8 583	58	1 000	6
North Korea	2 000	--	300	180
South Korea	921	--	386	--
Turkey	1 291	1 977	134	659
USSR	3 993 000	867 000	82 900	27 000
Other countries	5 368	353	1 488	74
Total	5 494 025	887 127	219 226	29 626

	Geological resources in 10^6 t.c.e.		Technically and economically recoverable reserves in 10^6 t.c.e.	
	h.c.	b.c.	h.c.	b.c.
Total world	7 724 834	2 400 430	492 472	143 892
	10 125 264		636 364	

Table 1: Coal Resources and Reserves, broken down according to Continents and Countries

Hard Coal (bituminous coal and anthracite)

Continent	Geological resources		Technically and economically recoverable reserves	
	in 10^6 t. c. e.	percentage	in 10^6 t. c. e.	percentage
Africa	172 714	2	34 033	7
America	1 308 541	17	126 839	26
Asia	5 494 025	71	219 226	44
Australia	213 890	3	18 164	4
Europe	535 664	7	94 210	19
Total	7 724 834	100	492 472	100

Brown Coal (subbituminous coal and lignite)

Continent	Geological resources		Technically and economically recoverable reserves	
	in 10^6 t. c. e.	percentage	in 10^6 t. c. e.	percentage
Africa	190	--	90	--
America	1 408 838	59	71 081	49
Asia	887 127	37	29 626	21
Australia	49 034	2	9 333	7
Europe	55 241	2	33 762	23
Total	2 400 430	100	143 992	100

Total

Hard Coal	7 724 834	76	492 472	77
Brown Coal	2 400 430	24	143 992	23
Gesamt	10 125 264	100	636 364	100

Table 2: The Distribution of World Coal Resources, grouped by Continents

Year	Geological Resources		Technically and economically recoverable reserves		Reserves expressed as percentage of Resources
	10^9 metr. t	10^9 t SKE	10^9 metr. t	10^9 t SKE	%
WEC 1974	10 754	8 603	591	473	5, 5
WEC 1976	11 505	9 045	713	560	6, 2
This Report 1977		10 125		636	6, 3

Table 3: Resource and Reserve Data Compared (WEC 1974, 1976, this report)
(Figures are not given in metric tons for this report because many countries replied, as requested, using calorific values as their basic unit)

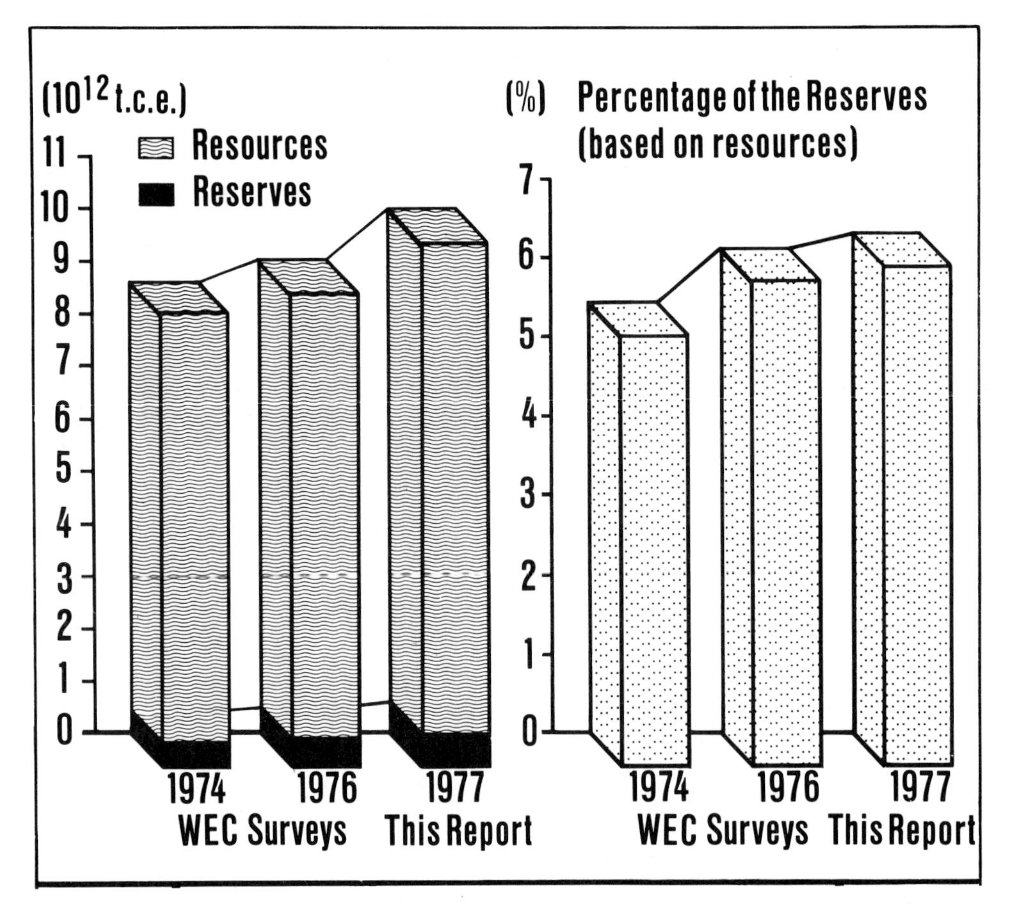

Fig. 1: Coal Resources and Reserves, and Ratio Reserves/Resources (%) for 1974, 1976, and for this Report

4.3.2. Evaluation of Figures for Resources and Reserves

It is assumed by a number of scientists that the figures for the resources given by the individual countries also contain occurrences, whose seam thicknesses are so slight, and whose depths are so great, that an economic exploitation can be practically excluded. It should, however, be noted in this connection that many countries have intensified their exploration activities, and in the assessment of their occurrences and resources are evidently applying stricter criteria than was earlier the case. Practically all the data given were accompanied with particulars on seam thickness and depth, which in part should be regarded as realistic under the viewpoint of extracticability. In many important countries, the data given on the resources are moreover clearly below the data that apply to occurrences, so that the figure shown in Table 3, i.e. approx. $10,000 \times 10^9$ t.c.e., are within an order of magnitude that can be regarded as realistic for those resources that may some day of economic interest for the population of the world. In view of the insufficient degree of exploration in many areas of the world, this figure might represent a lower limit. This also seems to be indicated by the rising tendency in reserves and resources during the past few years.

With regard to the reserves that are considered today as technically and economically recoverable, their share of approx. 6 % in the resources is very low. From these facts it can be inferred that, in general, strict criteria have been applied in assessing and evaluating this category. Many countries, such as the USA, for example, report relatively shallow depths and large seam thicknesses in consequence of the energy-economic peripheral conditions prevailing there. The exploitation factor, i.e. the ratio of the in-place reserves to the actually recoverable reserves is, in part, (e.g. in the USA), quoted at only 50 %, a value that must likewise be regarded as low. Basically this figure might apply, to the room and pillar mining; for longwall mining substantially higher exploitation factor values can be reckoned with.

The conclusion can be drawn from these facts that the category comprising the present technically and economically recoverable reserves has still a further considerable enlargement potential. Even slight increases in the energy price level of oil and natural gas may well lead to an appreciable increase in these reserves. In this connection it might be appropriate to consider also new technical developments. New developments in mining technology (e.g. hydraulic mining) which aim at greater mechanisation and automisation, and which also make it possible to work at greater depths, could possibly increase substantially the exploitation factor, and at the same time transform resources into additional reserves. This, as well, will depend on peripheral economic conditions. A further part of the resources could be upgraded to reserves, firstly, if, for example, the underground gasification of coal could be rendered more efficient and economic,(this technique has been worked on for some decades past); secondly, or if it were possible to carry out successfully a microbiological conversion to gaseous or liquid products ,(however, many experts currently doubt the feasibility of this technique).

4.3.3. The Geographical Distribution of the Coal Resources and Reserves

The geographical distribution of the coal resources and reserves are graphically represented in Figs 2 and 3.

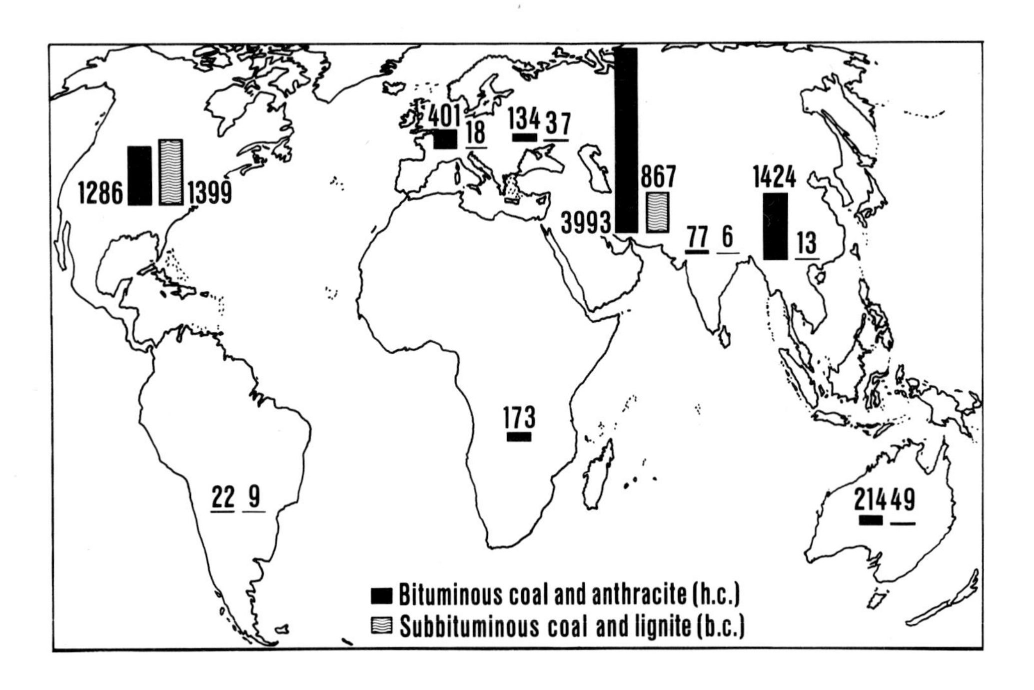

Fig. 2: Distribution of Coal Resources in the World (10^9 t.c.e.)

It is a conspicuous fact that the major part of the coal resources and reserves are concentrated in the northern hemisphere, with a number of countries ranking top of the list, viz.:

The Soviet Union
The United States of America
The People's Republic of China and
a number of European countries.

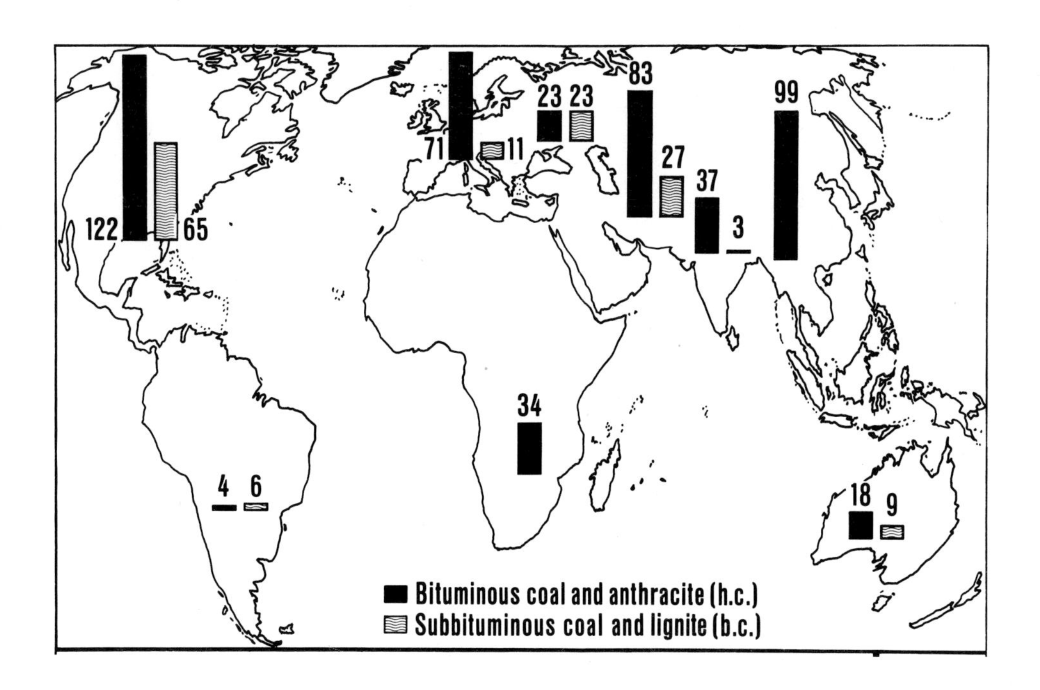

Fig. 3: Geographical Distribution of the Technically and Economically Recoverable Reserves (10^9 t. c. e.)

With regards to the geological hard coal resources, 85 % of these are accounted for by the Soviet Union, the United States of America, and the People's Republic of China; while the corresponding figure for the geological brown coal resources is 94 %. The corresponding figure for the technically and economically recoverable reserves are 60 % and 64 % respectively.

In the southern hemisphere the volume of the resources and reserves appear to be substantially lower. Up to now only three countries have important coal deposits, namely:

Australia,
The Republic of South Africa and
Botswana.

The geological resources of hard coal and brown coal in these countries account for 5 % of the total world resources; the figure for the technically and economically recoverable reserves being 9 %.

This different distribution may be due to the fact that the land mass of the southern hemisphere is smaller and that exploration has not advanced so far as yet in the countries of the southern hemisphere. On the other hand, however, it would appear that the geological conditions prevailing in some areas largely exclude to a large extent the presence of major coal deposits. Despite this, the fact should be reckoned with that an increase in exploration activity might well result in the discovery of further resources and reserves. The figure of $1,000 \times 10^9$ tons quoted as an maximum by Botswana, for instance, in its completed questionnaire, but not included in this report owing to still existing uncertainties, as well as the further successful exploration projects in Indonesia and Australia would appear to confirm this assumption.

5. The Availability of the Coal Reserves

5.1. Preliminary Remarks and Conduct of the Investigation

The investigation of the coal resources and coal reserves has shown that the deposits in many countries contain large quantities of reserves, which have even a still greater expansion potential, so that, in terms of quantities, these reserves could also cope with a sharply rising demand over a long period. These reserves, however, will only constitute an assured basis of supplies once they have become available on the energy market in good time, and in sufficient quantities. As far as a sufficient regional availability is concerned, it is of course first and foremost the figures for the future planned production of the individual countries that are of interest. For a global availability it is naturally the planned export rates of the individual countries that are of primary importance.

The realisation of these targets and estimate figures, and the timely availability of the coal reserves, is subject to obstacles of various degrees of seriousness, which have to be overcome. These obstacles occur from the exploration of the deposit via the actual extracting of the coal right through to the consumption of the coal. In this connection the following criteria for the availability were subjected to a closer examination:

1. The conditions prevailing in the deposit
2. Infrastructure
3. The coal trade and future export plans
4. Environmental impact
5. Coal quality
6. Coal conversion processes

7. Manpower and capital
8. Domestic coal supplies

This scheme was employed for the following countries:

- Australia
- Canada
- Federal Republic of Germany
- Great Britain
- India
- Japan
- People's Republic of China
- Poland
- Republic of South Africa
- USA
- USSR

These countries represent 92 % of the technically and economically recoverable coal reserves.

In addition to the above, the other countries of Europe and the countries of South America were studied together as separate groups. The results are set out in the next section.

5.2. Estimated Possible Development of Coal Production in the Main Coal-Producing Countries

5.2.1. Estimated Future Coal Production

The estimated figure of the future world coal production is shown in Table 4 for the main coal-producing countries, while those for the other countries are listed in Table 5. (Japan was included in table 4 because of its highly important place in the world coal trade.) Fig. 4 gives a graphic presentation of the future trend on coal production figures for the main coal producing countries. Some countries were not yet in the position to give figures for the year 2000, and only three main coal producing countries were able to give data for the year 2020 on their completed questionnaires. In a number of cases reference was made to the literature, and in some cases we made our own extrapolations based on the reserve data and the development of future coal production foreseeable. As a result, the following figures were obtained:

production rate	1985:	3.9×10^9 t.c.e.
	2000:	5.8×10^9 t.c.e.
	2020:	8.8×10^9 t.c.e.

Country	Coal production in 10^6 t.c.e.				Rate of increase %/a			
	1975	1985	2000	2020	1975 to 1985	1985 to 2000	2000 to 2020	1975 to 2020
Australia	69	150	300	400	8.1	4.7	1.4	4.0
Canada	23	35	115	200	4.3	8.2	6.1	4.9
China (P.R.)	349	725	1200	1800	7.6	3.4	2.0	3.7
Fed. Rep. of Germany	126	129	145	155	<0.1	0.8	0.3	0.5
India	73	135	235	500	6.3	3.8	3.8	4.4
Japan	19	20	20	20	<0.1	--	--	--
Poland	181	258	300	320	3.6	1.0	0.3	1.3
Republic of South Africa	69	119	233	300	5.6	4.6	1.3	3.3
United Kingdom	129	137	173	200	0.6	1.6	0.7	1.0
USA	581	842	1340	2400	3.8	3.1	2.9	3.2
USSR	614	851	1100	1800	3.3	1.7	2.5	2.4
Total	2233	3401	5161	8095	4.3	2.8	2.3	2.9

Table 4: Estimated Production of the Main Coal Producing Countries for 1985, 2000 and 2020

Country	Coal production in 10^6 t.c.e.				Rate of increase %/a			
	1975	1985	2000	2020	1975 to 1985	1985 to 2000	2000 to 2020	1975 to 2020
Argentina	0, 7	3, 0	6	8	15, 6	4, 7	1, 4	5, 5
Belgium	7, 0	7, 2	7	7	-	-	-	-
Botswana	-	5, 5	11	16	-	4, 7	1, 9	-
Brazil	2, 5	7, 5	15	40	11, 6	11, 6	5, 0	6, 4
Bulgaria	13, 6	18, 7	30	35	3, 2	3, 2	1, 4	2, 1
Chile	1, 5	2	6	8	2, 9	7, 6	1, 4	3, 8
Colombia	3, 6	8	15	25	8, 3	4, 3	2, 6	4, 4
Czechoslovakia	80, 0	93	100	110	1, 5	0, 5	0, 5	0, 7
France	23, 4	14	14	14	./. 5, 0	-	-	./.1, 1
German Democr. Republic	74, 6	80	90	100	0, 7	0, 7	0, 5	0, 6
Greece	6, 0	14	18	20	8, 8	1, 7	0, 5	2, 7
Hungary	10	24	24	25	9, 1	-	0, 2	1, 3
Indonesia	0, 3	2	13	18	25, 9	10, 3	1, 6	9, 5
Mexico	7, 1	20	42	45	10, 9	5, 1	0, 3	4, 2
North Korea	34, 0	36	40	50	0, 6	0, 7	1, 1	0, 8
Romania	13, 0	25	35	40	6, 7	2, 3	0, 7	2, 5
South Korea	18, 0	24	20	20	2, 9	./. 1, 2	-	0, 2
Spain	12, 3	25	23	25	7, 3	./. 0, 5	0, 4	1, 6
Turkey	13	21	30	35	4, 9	2, 4	0, 8	2, 2
Venezuela	0, 2	5, 5	6	10	39, 3	0, 6	2, 6	9, 1
Yugoslavia	18, 1	21	40	45	1, 5	4, 4	0, 6	2, 0
Other countries	21	26	34	55	2, 2	1, 8	2, 4	2, 2
Total	360	483	619	751	3, 0	1, 7	1, 0	1, 6
Main coal producing countries	2233	3401	5161	8095	4, 3	2, 8	2, 3	2, 9
Total world production	2593	3884	5780	8846	4, 1	2,7	2,1	2, 7

Table 5: Estimated Coal Production of the Other Countries and World Coal Production for 1985, 2000 and 2020

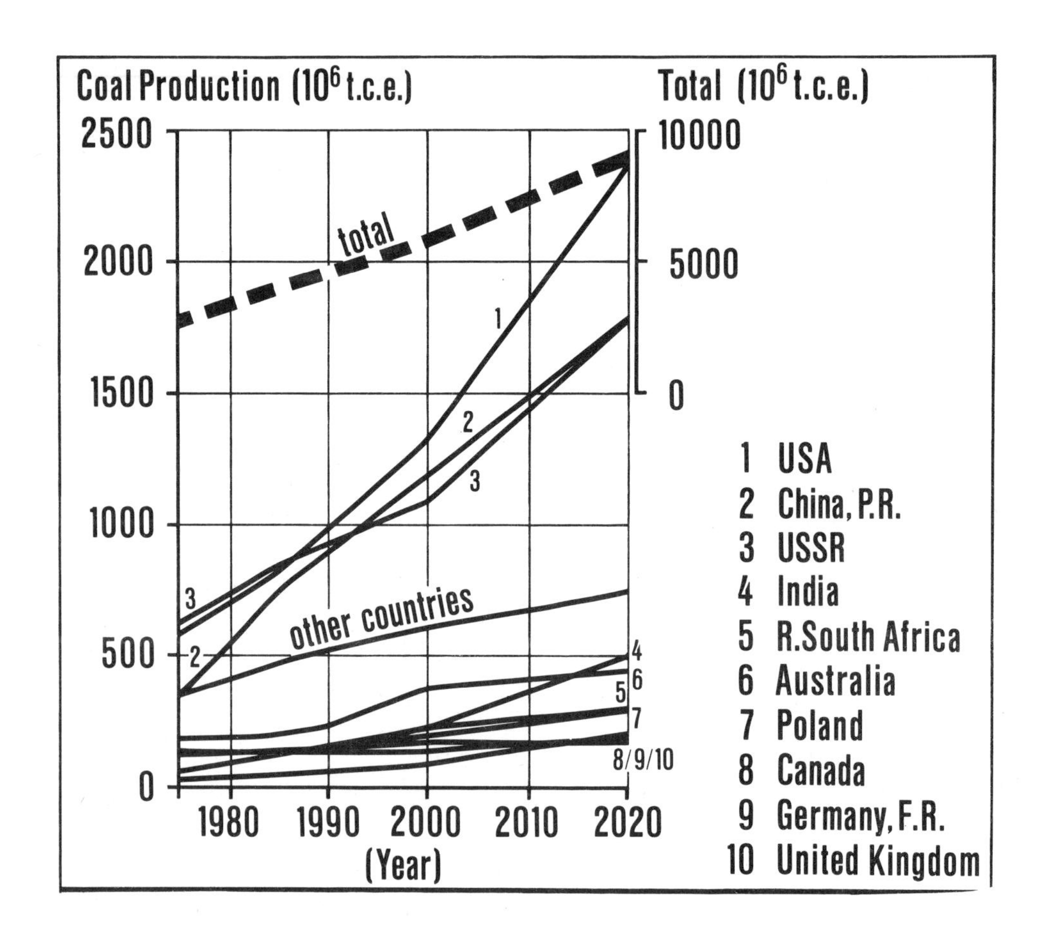

Fig. 4: Survey of the Future Trend in Production Figures for the Main Coal-Producing Countries

5.2.2. Evaluation of the Production Data

The production figures found out, especially those for the years 2000 and 2020, must be regarded as maximal values that can be attained with the aid of the currently initiated measures. The sum of these figures, which yielded the above results, would therefore have to be regarded also as an upper limit that could be achieved, based on today's assessments, with the measures presently being implemented.

This conservative evaluation results from the fact that the future coal production,and also coal consumption,may be faced with a number of obstacles (in connection with which there is admittedly a fair degree of probability that, as some time in the future, they will occur in a more or less drastic form, though this is not quantifiable today). The results of this investigation are set out as a brief summary in the following section.

5.2.3. Possible Bottlenecks in the Sufficient Availability of Coal

In most coal-producing countries, increased activities are being carried out, aimed at prospecting and developing new deposits. A significant increase in production can only be realized in many of the coal-producing countries by relocating the principal mining centres to areas that are less developed industrially. In the USA, for example, one observes a transfer of the coal-mining operations from the eastern territories to the large deposits located in the Midwest and the western part of the country (Fig. 5).

Fig. 5: Future Relocation of Coal Mining, USA

The deposits in the European part of the USSR appear to have a limited lifetime only. Here, too, there might well be a gradual transfer of the principal coal-mining operations to the coal deposits east of the Urals and Siberia in the course of the next decades (Fig. 6).

Fig. 6: Future Relocation of Coal Mining, USSR

In this context, it might well be, (in many other countries and not just in the USA and USSR), that the long lead times involved might be problematic: these long lead times can last from 5 to 15 years in present conditions, and are the result of the work needed from the establishment of new mines to the time of full production. Furthermore, in many countries, there is the added necessity of developping or expanding the requisite infrastructure and transport capacity, while other situations require the building of suitable conversion plants.

Measures such as these are being initiated in many areas, but necessitate high capital investment in addition to the long lead times. However, experience seems to show that it is only possible to bring this about to an adequate extent, if there are good prospects of the rentability of the initial capital investment in question within a reasonable period. The point in time when a rapidly increasing worldwide growth in demand will start to develop is uncertain, a fact which, needless to say, compels the energy industry to proceed cautiously with their forward planning, to a large extent. Capital investment funds might be deployed at a time of shortage on the oil and natural gas market, but this might well be too late, as the productive effects of these measures would only become manifest at a relatively later date, (because of the long lead times that are specific to the coal industry). These factors might therefore constitute a serious obstacle to the coal reserves becoming available at the right time. One should ask the question, therefore, whether, in the near future, additional energy-political measures should not be taken in precisely this area, so that the energy industry might be provided, in good time, with sufficient incentives and assistance for expanding the production and transportation capacities.

Any drastic raising of the coal output is conditional on the provision of qualified skilled personnel. According to the information currently available, this demand might well present many producing countries with considerable problems. In particular, some countries will have to have recourse to underground mining, and, because of the conditions of the deposits, will have to transfer their coal-mining operations to regions that have hitherto hardly been used for industrial purposes, and which have a low population density.

In the area of environmental protection, increasing consideration will have to be given by the mine operators to the recultivation of the worked-out opencast areas. In the case of the consumer we have to take into account, that the existing market structure and the increasing demand for comfort by the general population will lead in the future to the demand for fuels that are easy to handle, (mainly in the form of electricity, as well as gaseous and liquid hydrocarbons). As far as the future is concerned, these will have to comply with the increasing demands of the environmental regulations. Research and development activities directed towards this end are being actively pursued in many countries. It appears, however, that these efforts will have to be intensified still further, if the processes being evolved can be placed on the market in good time. In the fields of power and heat generation, the fluidized bed combustion process, among others, appears to be attracting particular interest, since it is able to reduce considerably the emissions of sulphur dioxide and nitric oxide by relatively simple means. It is also suitable for the use of coals with high sulphur and ash contents, so that, if it is successfully developed, the operation basis of the coal conversion plants will be expanded and the availability of coal increased.

More or less the same applies to the coke production sector. In this field, as a result of a successful development of the particularly environment-compatible process of continuous coking, the coking-coal category can be extended to include coals that were hitherto not cokable.

Also in the coal gasification and liquifaction sectors there are processes in the development stage that also promise to have a higher environment-compatibility factor. These processes are either improvements on past techniques, or completely new ones. Whether such processes will be available in good time or not, will depend (among other things) on sufficient funds being made available for their development.

As far as the basically non-toxic carbon dioxide (CO_2) is concerned, (which is liberated on the combustion of fossil fuels), there is still no consensus of opinion today on whether the relatively slight increase in the CO_2 content during the past few decades is actually attributable to combustion processes, since in nature there are still greater sources of CO_2. Nor is there any certain knowledge today on whether a higher CO_2 content in the atmosphere can have an influence on the climate.

The hitherto projected production figures appear to be sufficient in many countries for increasing the regional availability. The overall problems involved in the future availability of coal can only be rendered clear by including global aspects in the considerations. Global availability will depend to a decisive degree on whether and to what an extent the main coal-producing countries of the future are planning to provide coal for export.

In this respect, primary importance must be accorded to this aspect. The results achieved by this investigation are summarized in the following section.

5.2.4. Global Availability (Exportavailability)

The world trade volume in coal, consisting almost entirely of bituminous coal, currently amounts to approx. 200×10^6 t.c.e. per annum. The oversea coal trade accounts for around 100×10^6 t.c.e. In terms of calorific value, this is equivalent to only about 5 % of the world overseas trade in oil.

Table 6 reviews the planned and estimated export figures. For reasons of facilitating comparability, the production figures have been included in the Table. The percentages of the export rates in the production figures, likewise shown in the Table, are listed in Fig. 7. In this respect, too, a number of countries were not in a position to state their export rates up to the year 2020, so that in part, we had to resort to data derived from the literature and to our own extrapolations. The investigation yielded the following export rates:

1975:	199×10^6 t.c.e.	(7.7 % of total output)
1985:	303×10^6 t.c.e.	(7.8 % of total output)
2000:	582×10^6 t.c.e.	(10.1 % of total output)
2020:	788×10^6 t.c.e.	(8.9 % of total output)

Country	1975			1985			2000			2020		
	pro-duction	export	export (%)	pro-duction	export	export (%)	pro-duction	export	export (%)	pro-duction	export	export (%)
Australia	69	29	42	150	60	40	300	180	60	400	240	60
Canada	23	12	52	35	15	43	115	40	35	200	65	32
China (People's Rep.)	349	3	1	725	7	1	1200	30	2	1800	50	3
Fed. Rep. of Germany	126	23	18	129	25	19	145	30	21	155	30	19
India	73	-	-	135	7	6	235	13	7	500	32	6
Japan	19	-	-	20	-	-	20	-	-	20	-	-
Poland	181	39	21	258	45	17	300	50	17	320	50	16
South Africa (Rep.)	69	3	4	119	23	19	233	55	24	300	60	20
United Kingdom	129	2	2	137	10	7	173	10	6	200	10	5
USA	581	60	10	842	68	8	1340	90	7	2400	145	6
USSR	614	26	4	851	37	4	1100	50	5	1800	60	3
Other countries	360	2	1	483	6	1	619	34	6	751	46	6
Total	2 593	199	7,7	3884	303	7.8	5780	582	10.1	8846	788	8.9

Table 6: Data of production and export of the main coal producing countries (10^6 t.c.e.)

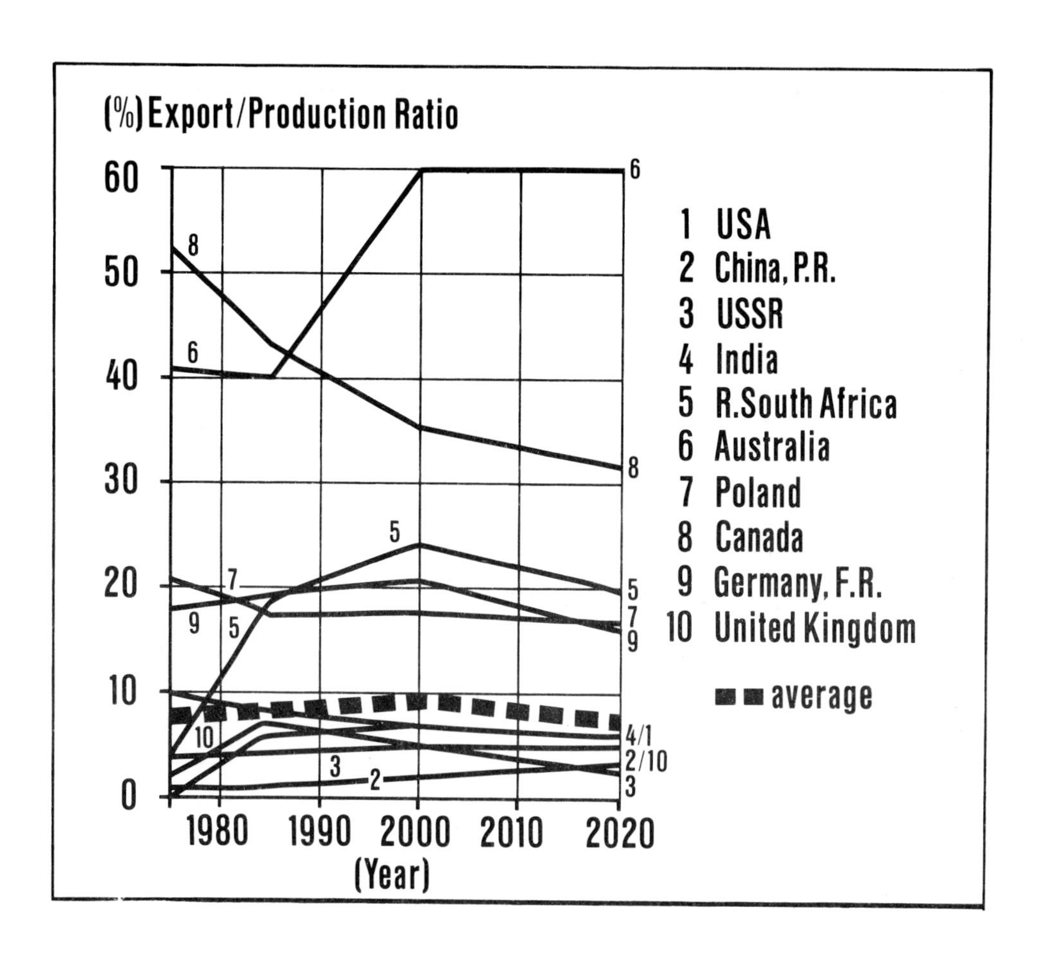

Fig. 7: Development of the Export/Production Ratio (%) of Main Coal Producing Countries

Hence, the export percentage ascertained for the year 2000 lies - in terms of calorific value - far below 25 % of today's world export volume in oil and natural gas.

This result appears to show that the main coal-producing countries are today still orientating their coal production to their own consumption, and that it will hardly be possible to build up an adequate trade in coal on the basis of this comparatively low export rate. Fig. 8 shows as a summary, the trend of production and export until the year 2020.

It should, of course, be emphasized that these results are based on the presently available information and on the currently existing and projected measures. Significant changes on the energy market, that result in high world coal prices, may naturally constitute an increased incentive for a higher coal production, and lead to increased exports. The abovementioned long lead times make it nevertheless appear doubtful

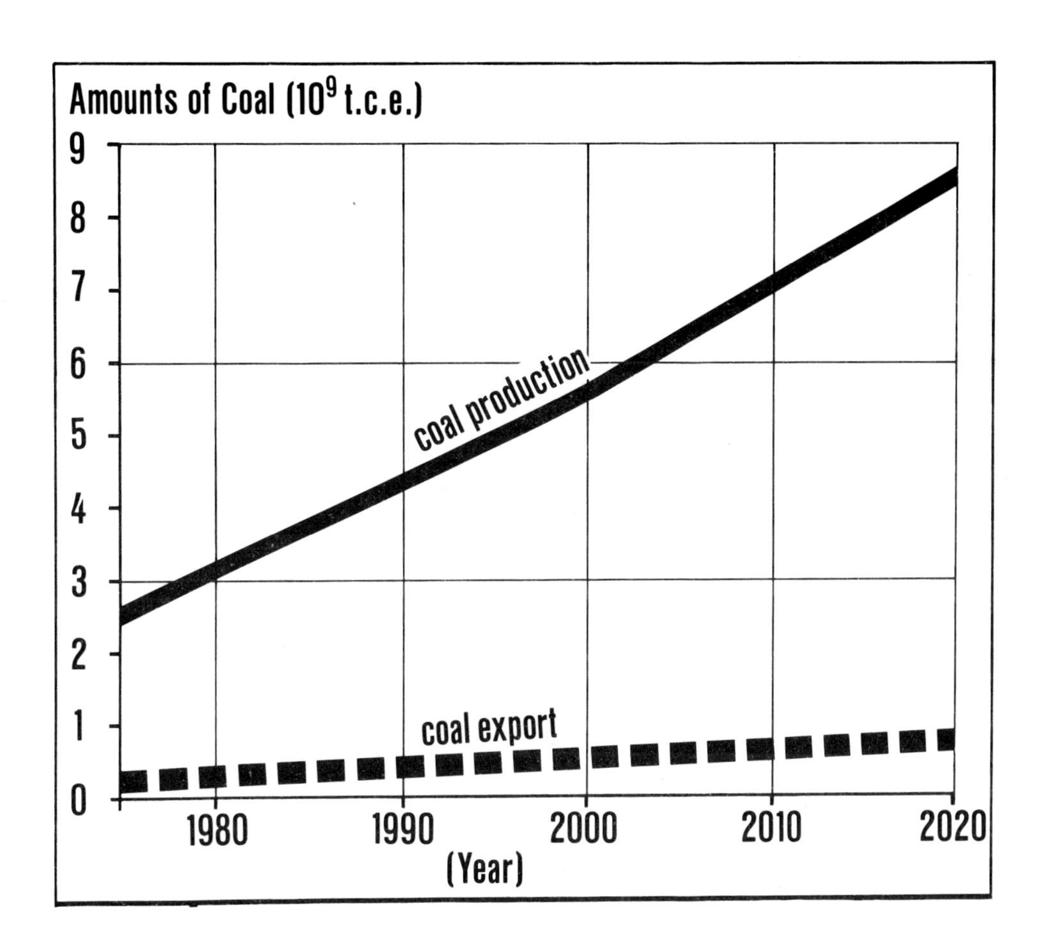

Fig. 8: Development of Coal Production and Coal Export - World

that the gap in demand, which could well occur in the near future, can be closed with certainty. If this is to be achieved, additional preparatory measures will be required, which would have to be initiated at the right time before the onset of any actual shortage.

6. Economic-Statistical Estimation of the Possible Trend in Production, based on Different Reserves Data until 2020

Finally, in addition to the results of the Study a tentative economical-statistical consideration should be introduced, in order to estimate whether the projected future production trends correlate with the extent of known reserves.

Such an approximation of the presumed future trends in coal production is shown in Fig. 9 in respect of various reserves figures, the basis used being a normal Gaussian curve.

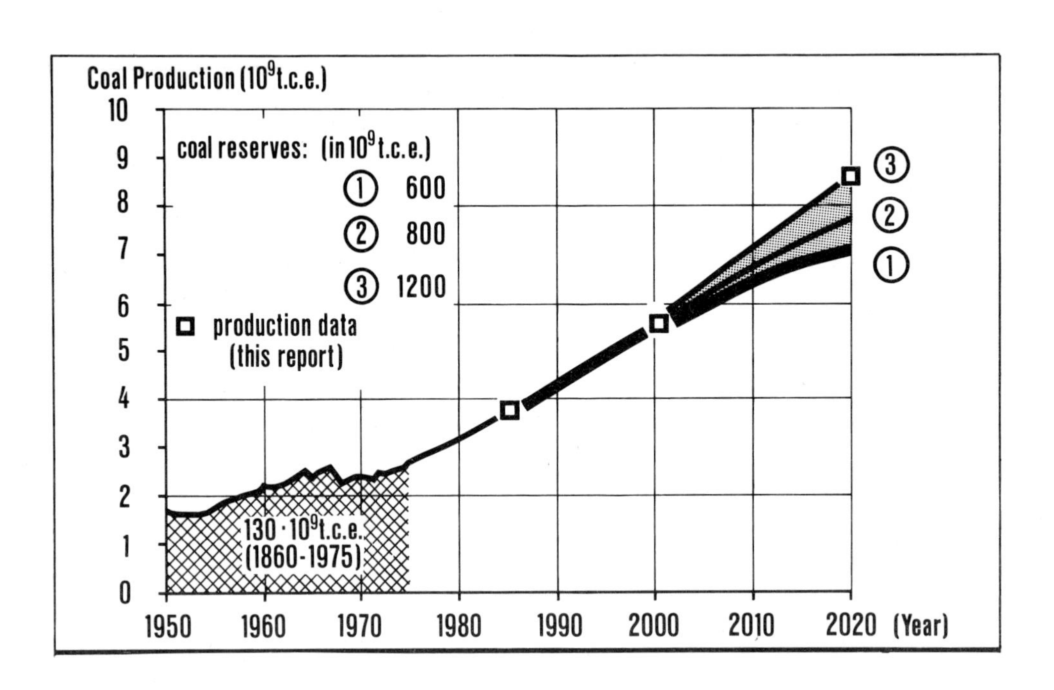

Fig. 9: Future Trend in World Coal Production, based on Different Amounts of Coal Reserves

The result of this is, that the curve for the reserve of 600 x 10^9 t.c.e. meets the production planned for 1985 and 2000, but not for the year 2020. This result shows that the reserves presently regarded as economically recoverable, and the production of 8.8 x 10^9 t.c.e. found in this report for the year 2020 are not in correlation. This means that this production figure cannot be met under normal circumstances and the reserves stated above. This production rate can be achieved with reserves of 1 200 x 10^9 t.c.e. or more. Since the potential of the coal reserves is so huge, and as we can expect the price of energy to rise on the market, it is very likely, that the figure of 600 x 10^9 t.c.e. will double. Because of this, from the standpoint of reserves, the production estimate for the year 2020 seems basically to be possible, and for a long time after that an increasing coal consumption could be a practical reality.

Also from this result one can see the high potential of coal for the ensuring of the energy supply of the future.

Taking into account the resource and reserve potential one can think that the meeting of the demand is basically possible. Nevertheless, it is doubtful that these reserves will be available in time. The reason is that the present efforts to increase the production of coal are not sufficient to meet the projected figures. It will be all the more necessary to intensify those measures including energy-political initiatives in this direction.

7. Possibilities of an Increased Coal Output[+)]

The facts and data presented in the Sections 1 to 6 are the result of thorough investigations carried out in close contact with the various coal mining countries. Therefore, these results represent the present state of knowledge in this field.

As mentioned in Section 5.4.2. already, the production rates and especially the export rates seem to be far too low to build up a sufficient world market for coal and to meet the import demand of those countries disposing only of small coal deposits. This ist true althemore as many developing countries have hardly any energy deposits of their own and as their respective growth rates will have to increase over-proportionally in order to raise the standard of living and to adjust to the more developed countries.

Nearly all coal producers consider today the missing economical incentive to be the main obstacle for a step up of coal output and of coal exports. If such an incentive could be established, e.g. by a regionally and worldwide followed energy and economic policy and also by new technical developments, the upper limit of the potential coal output would be determined only by technical conditions. During and after the Instanbul Conference, leading experts including those of the coal mining industry, expressed the view that a much more rapidly increasing coal output could be possible under technical conditions alone. This should be true especially for the large coal minig countries such as USA, USSR, Australia, the People's Republic of China and the Republic of South Africa. For USA, e.g. the growth rates and production figures given in Table 7 are considered to be possible under the conditions stated.

Growth Rates		Coal Output	
for period	% / a	in the year	10^6 t.c.e.
1975 - 1985	7.2	1975	581
1985 - 2000	3.8	1985	1.162
2000 - 2020	3.2	2000	2.018
1975 - 2020	4.3	2020	3.790

Table 7: Increase of coal output in USA considered technically feasible by experts

+) This section was added after several discussions during and after the World Energy Conference 1977 in Istanbul.

In case these optimistic assumptions are taken also for the above-mentioned other big coal mining countries and if the coal output of the remaining other countries is taken from the Tables 4 and 5 without change, the figures shown in Table 8 result.

Growth Rates for period	% / a	World Coal Output in the year	10^6 t.c.e.
1975 - 1985	5.7	1975	2.593
1985 - 2000	3.4	1985	4.503
2000 - 2020	2.9	2000	7.420
1975 - 2020	3.7	2020	13.061

Table 8: Maximum world output estimated by experts under technical viewpoints alone

According to these estimates, the average growth rates decrease continuously, but the absolute coal output increases substantially. The results in detail are as follows:

- the average growth rates for period 1975 to 2020 comes to 3.7 % per annum;

- an absolute volume of about 13 x 10^9 t.c.e. of coal output is considered to be technically feasible for 2020, a sufficient economical incentive provided. This figure is by about 5 x 10^9 tons higher than the figure resulting from the present plans of the individual mining countries.

This additional volume of 5 000 million t.c.e. amounting to about 40 % of the total world production, could be available for exports, the existence of a suitable infrastructure and of effective transport facilities provided.

HYDRAULIC RESOURCES

Ellis L. Armstrong
Department of Civil Engineering, University of Utah; *formerly* US Commissioner of Reclamation and 1972–1974 Chairman, US National Committee, World Energy Conference

TABLE OF CONTENTS

ABSTRACT

INTRODUCTION

DEVELOPMENT TO DATE

FACTORS AFFECTING DEVELOPMENT

Economics

Multipurpose Use of Water

Technology

Environment

Societal

Legal and Political

REQUIREMENTS FOR DEVELOPMENT

PROBABLE FUTURE DEVELOPMENT

ACKNOWLEDGEMENTS

CONSERVATION COMMISSION

WORLD ENERGY CONFERENCE

Executive Summary

HYDRAULIC RESOURCES

By

Ellis L. Armstrong

Engineering Consultant and Adjunct Professor
of Civil Engineering, University of Utah;
Formerly U.S. Commissioner of Reclamation and
1972-1974 Chairman, U.S. National Committee,
World Energy Conference

ABSTRACT

Hydraulic energy presently is providing about 23 percent of the world's electricity. However, it is of greater relative importance because it is a continually renewable, non-polluting energy resource; it is an integral part of optimum overall water resource utilization; it is an important part of large electric power systems because of its reliability and flexibility; it is a catalyst in economic improvement in developing countries; its economic justification is improving because of its "inflation proof" characteristics and its long life and low maintenance costs; and it provides important peaking-power capabilities.

The installed hydroelectric capacity at present is about 372,000 megawatts with a yearly production of 5.7 million terrajoules of energy (1.6 million gigawatt hours), which is approximately 16 percent of the total installed and installable capability as reported in the World Energy Conference 1976 Survey of Energy Resources. The Organization for Economic Cooperation and Development (OECD) Countries have developed 46 percent of their capacity, and the other countries, as an average, about 7 percent of their capacity. The development of hydroelectric resources is affected by economics, other uses and needs for water, environment, technology, and social factors as well as legal and political considerations. Large developments require sizeable multidiscipline organizations and extensive planning. On multinational rivers, international agreements must be reached for developments to proceed.

From extensive review of data and factors affecting future development, the author estimates the probable hydroelectric production by 2020 is likely to increase five times to about 28 million terrajoules, approximately 80 percent of the total developable hydroelectric resource reported in the 1976 WEC Survey. Providing adequate financing for this development will be a big problem, estimated to average 33 billion 1976 U.S. dollars per year over the 44 year period.

INTRODUCTION

Water is essential to all life. In addition, in its never ending cycle, it can be harnessed to provide energy for man's use. Water flowing under pressure is converted first into mechanical energy and then into electricity.

Hydraulic energy presently is generating 23 percent of the world's electricity. It will increase in importance for the reasons discussed in the following paragraphs.

1. It is a resource continually renewable by the energy of the sun which creates and sustains the hydrological cycle. Water is evaporated from the oceans and from where it occurs on land areas, is carried by air movements over the global surfaces and condenses as temperature and pressure conditions change. Part of the precipitation falls on upland areas to create the streams, rivers, and lakes of the world, through which the water flows back to the oceans. Man's efforts to improve the amount of useable hydraulic energy include measures to level out the river flows by creating reservoirs to store flood waters for later controlled releases. Ground water recharging, such as through soil and plant management, increases rainfall percolation into the ground and thus sustains dry season flows. Increasing rainfall on higher locations by scientifically controlled cloud seeding programs, thereby increasing streamflow available for use, has promise in special locations. This is possible as nature, without help from man, is an inefficient precipitator; 90 percent or more of the moisture in the clouds passes on across the land areas and back to the oceans.

2. Hydraulic energy is non-polluting. Whereas fossil and nuclear fuel uses produce polluting by-products along with heat and mechanical energy, the energy of falling water is virtually pollution-free and no heat is released. Reservoirs can create problems, but with balanced management, can also greatly enhance the natural environment. Floods are controlled and low late seasons flows are or can be increased.

3. Hydroelectric energy can be, and usually is, an important part of multipurpose utilization of water resources. Water for irrigation of crop lands, municipal uses, navigation, pollution control, recreation, and fish and wildlife usually requires control of river flows, which control can also with proper planning, provide hydroelectric energy without adversely affecting the other uses. Some tradeoffs usually will be necessary to optimize the overall benefits, but hydroelectric generation can often make comprehensive utilization of water resources economically possible. The requirement is for a fully integrated water resource development program with careful attention to all needs and potential uses.

4. Hydroelectric energy in small quantities can be made available at remote areas of developing countries and can be a catalyst in developing other resources and creating opportunities for improving human conditions. This has proven to be so in the past, and probably will be even more important in the future. While the world's largest power generating installations are hydroelectric, increasing attention is being given to small installations of a few kilowatts. For instance, while the Chinese have some large hydroelectric installations, it is reported that they have constructed about 50,000 hydroelectric plants within the past decade with an average capacity of 35 kilowatts. The development and greater use of facilities such as the bulb-type self contained turbine-generator, with remote control, is expected to further expand the use of small hydroelectric installations to utilize the natural flow of rivers.

5. The reliability and flexibility of operation, including the fast startup and shutdown time in rapid response to changes in demand, make hydroelectric installations an especially useful part of a large electric power system, greatly increasing its performance and efficiency. They provide spinning reserve for emergencies, as well as convenient, economical and effective peaking power.

6. The long life and low operating costs, in view of the recent increases and further increases expected in fossil fuel costs, makes hydroelectric energy much more attractive economically. Several times as much capital investment per kilowatt as in the immediate past can now be economically justified in hydroelectric installations.

7. The technology and efficiency of hydroelectric production is well developed and proven. Turbines with efficiency as high as 95 percent or more are now available. Units as large as 700,000 kilowatts, probably at this size close to the upper limit in economic efficiency, are in operation. With mass production and marketing applied to small hydroelectric units, 10 to 50 kilowatts or so, as is likely especially in the developing countries, capital costs can be reduced.

8. The improvements made in technology in recent years make possible significant increases in the generating capability of existing plants, up to 10 percent or more, with minimum effort and favorable costs. This may be accomplished by rewinding of generators and improving the turbines. For instance, recent rewinding of the generators at Shasta Dam in the United States, increased production about 15 percent. Further, additional generation can be provided by extending existing plants to take advantage of the full potential of the site considering changes in hydrology of the watersheds, the needs for peaking power, and the increased value of hydroelectric generation as a part of a large electric power system including large capacity coal-fired and nuclear plants. In some instances rebuilding of existing facilities is warranted.

9. The flexibility and dependability of hydroelectric generation makes energy storage by means of hydraulic pumped storage systems, the most economical and trouble-free method available to date for large scale use. While four kilowatt hours of energy input is required for each three kilowatt hours of energy utilized, the input is low cost energy and the output is high value capacity for peaking loads. In a large electric power system, where the only alternative may be the utilization of old and relatively inefficient thermal units to meet peak loads, the use of hydroelectric pumped storage can result in overall saving in energy. An all underground pumped storage system has great potential, especially in areas of flat terrain.

In view of these advantages, it is expected that all potential hydroelectric power developments, including upgrading and modification or rebuilding of existing plants, will receive new and intensive consideration. The potentials, especially in the developing countries, are great. While the most favorable sites have been developed in the industrialized countries, a large potential for better utilization of the hydroelectric resources remain.

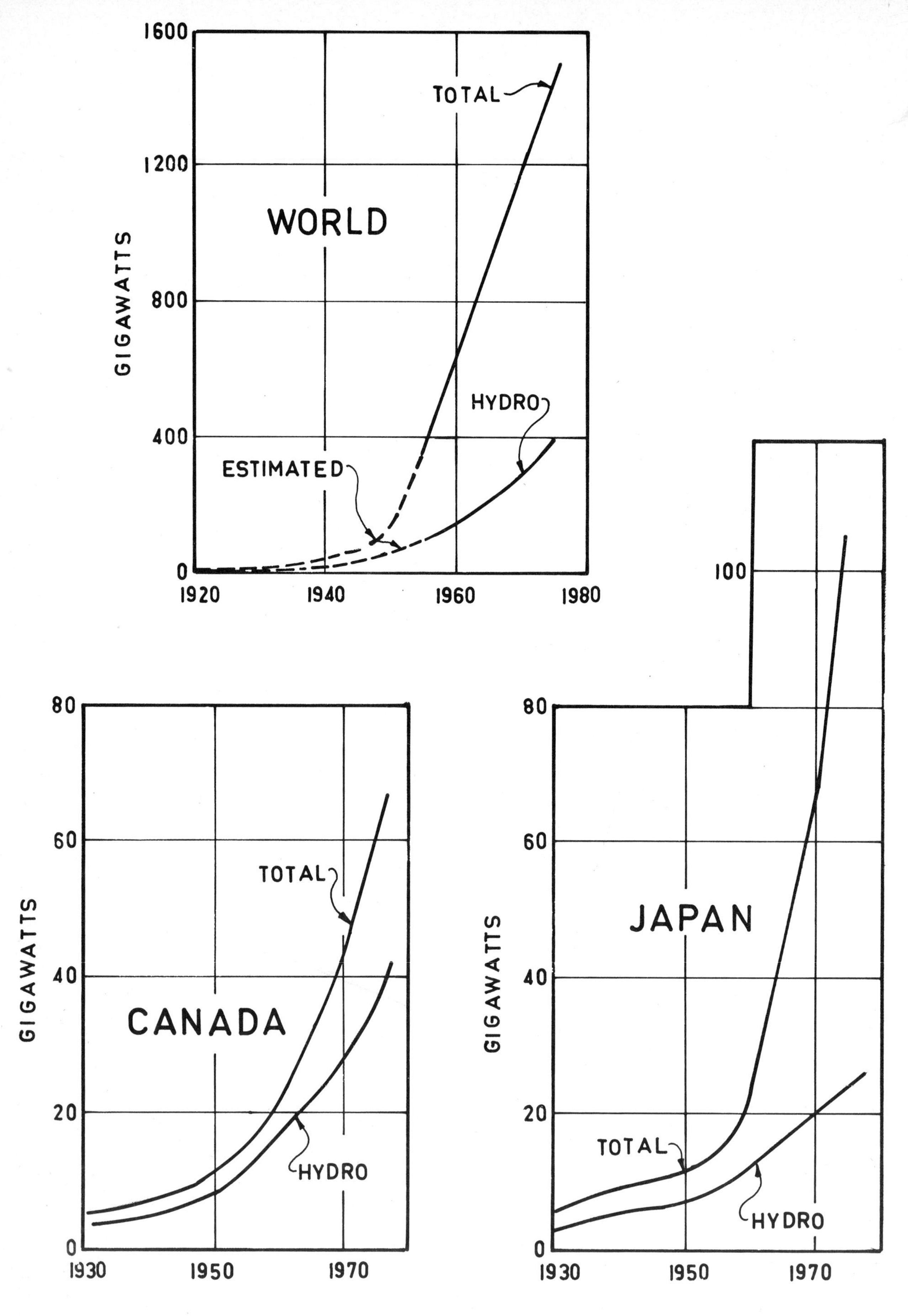

ELECTRIC GENERATING CAPACITY

FIGURE 1

DEVELOPMENT TO DATE

Man's earliest extensive use of energy, other than muscle power of man and animals, was that derived from flowing water. The use of waterwheels of various types extend back to the early civilizations. The size and efficiency of waterwheels increased over the centuries, and in the nineteenth century, water-powered mills of various types ushered in the beginning of the industrial age. The peak of this early water power development phase was reached about the middle of the nineteenth century because favorable mill sites, within the reach of the mechanical transmission of power, were limited. Furthermore, at that time, the more flexible steam engines were improving in economy and dependability.

With the advent of electricity in the 1880's, and with alternating current technology making transmission of electric energy more economical, the development of hydroelectric energy was well underway by the beginning of the twentieth century. Developments were rapid and by the 1930's, projects such as the 1.3 million kilowatt powerhouse at Hoover Dam in the United States were completed. Large hydroelectric installations such as this increased the utilization of energy in the industrialized nations and programs to utilize the large hydroelectric potentials were pushed ahead. As demands for electrical energy grew, even the vast hydroelectric potentials of areas such as the Columbia River Basin in the United States, by the 1950's and 1960's, required large installations of thermal power generation to meet the growth in demands.

Figure 1 illustrates the growth of total electric generating capacity compared to that of hydroelectric capacity. The data for the world totals is from the United Nations Compilation of World Energy Supplies from 1955 to 1974. These same relationships for Canada and for Japan are also illustrated.

The information from the 1976 World Energy Conference Survey of Energy Resources (as shown in Figure 2) indicates there is a total potential from hydraulic resources of 2.2 million megawatts of installed and installable generating capacity with a potential annual energy production of 34.9 million terrajoules (9.7 million megawatt hours). These figures were determined from the capacities at the various sites on a river of the plants installed and those which probably would be installed to utilize the energy resources of the site. Economics in some instances limited the installable figures. To produce 34.9 million terrajoules of electrical energy would require the burning of about 14.6 billion barrels of oil, or 40 million barrels of oil per day on a yearly basis, in oil-fired thermal power stations.

The operating hydroelectric capacity at present is about 372,000 megawatts with a yearly production of 5.7 million TJ (1.6 million GWH). This is approximately 16 percent of the total reported installed and installable potential. By 2020 the author has estimated the development will be about 80 percent of the total. To produce 5.7 million TJ of electricity in oil-fired thermal plants would require the burning of 2.5 billion barrels of oil, or 6.5 million barrels of oil per day on a yearly basis.

Figure 2 shows the overall potential of the various areas of the world: Asia, 28%; South America, 20%; Africa, 16%; North America, 16%; USSR, 11%; Europe, 7%; and Oceania, 2%. The vertical shows the percentages in each

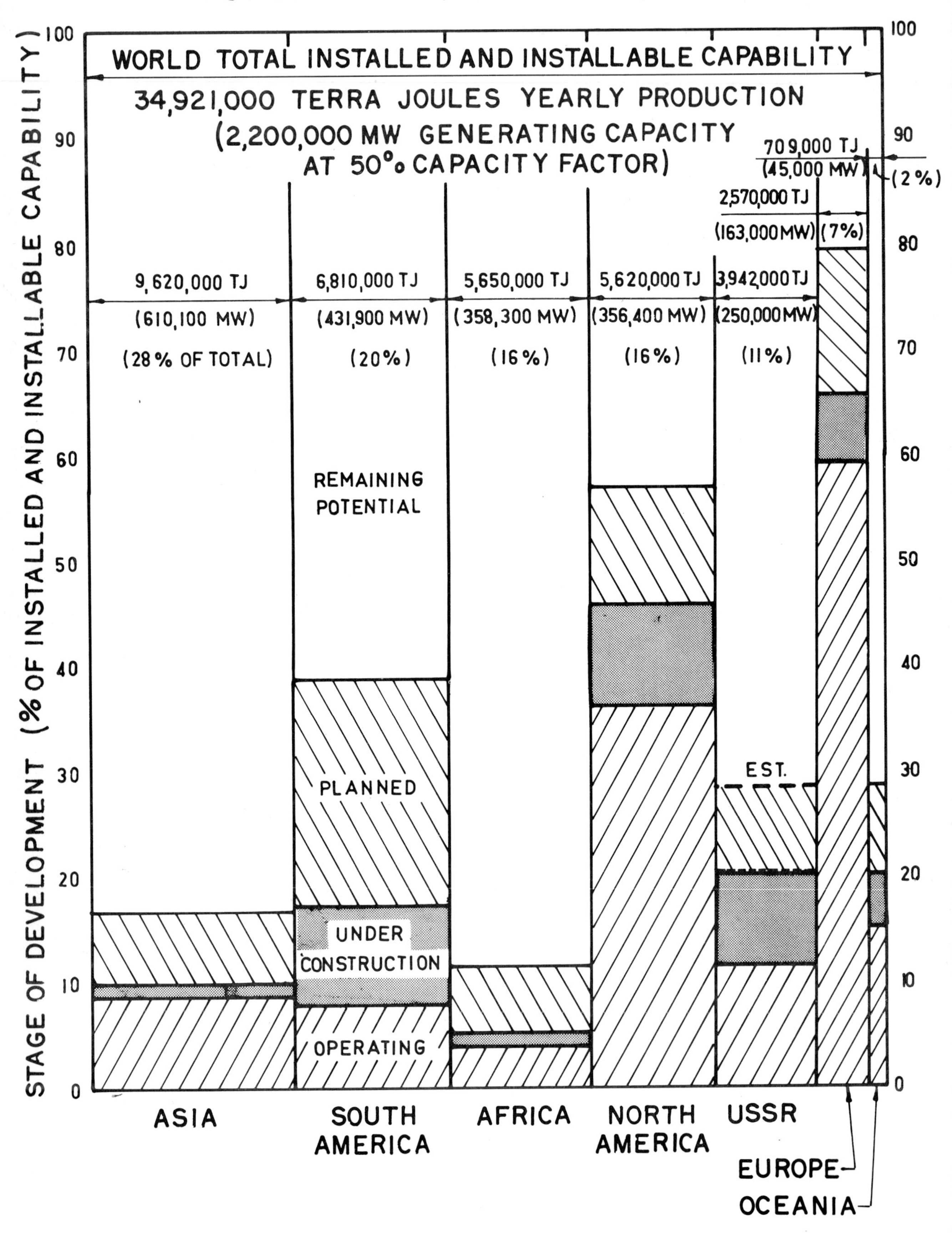

WORLD HYDRAULIC RESOURCES
WORLD TOTAL INSTALLED AND INSTALLABLE CAPABILITY
34,921,000 TERRA JOULES YEARLY PRODUCTION
(2,200,000 MW GENERATING CAPACITY
AT 50% CAPACITY FACTOR)
STAGE OF DEVELOPMENT (% OF INSTALLED AND INSTALLABLE CAPABILITY)
9,620,000 TJ
(610,100 MW)
(28% OF TOTAL)
6,810,000 TJ
(431,900 MW)
(20%)
5,650,000 TJ
(358,300 MW)
(16%)
5,620,000 TJ
(356,400 MW)
(16%)
3,942,000 TJ
(250,000 MW)
(11%)
2,570,000 TJ
(163,000 MW)
(7%)
709,000 TJ
(45,000 MW)
(2%)
REMAINING POTENTIAL
PLANNED
UNDER CONSTRUCTION
OPERATING
EST.
ASIA
SOUTH AMERICA
AFRICA
NORTH AMERICA
USSR
EUROPE
OCEANIA

FIGURE 2

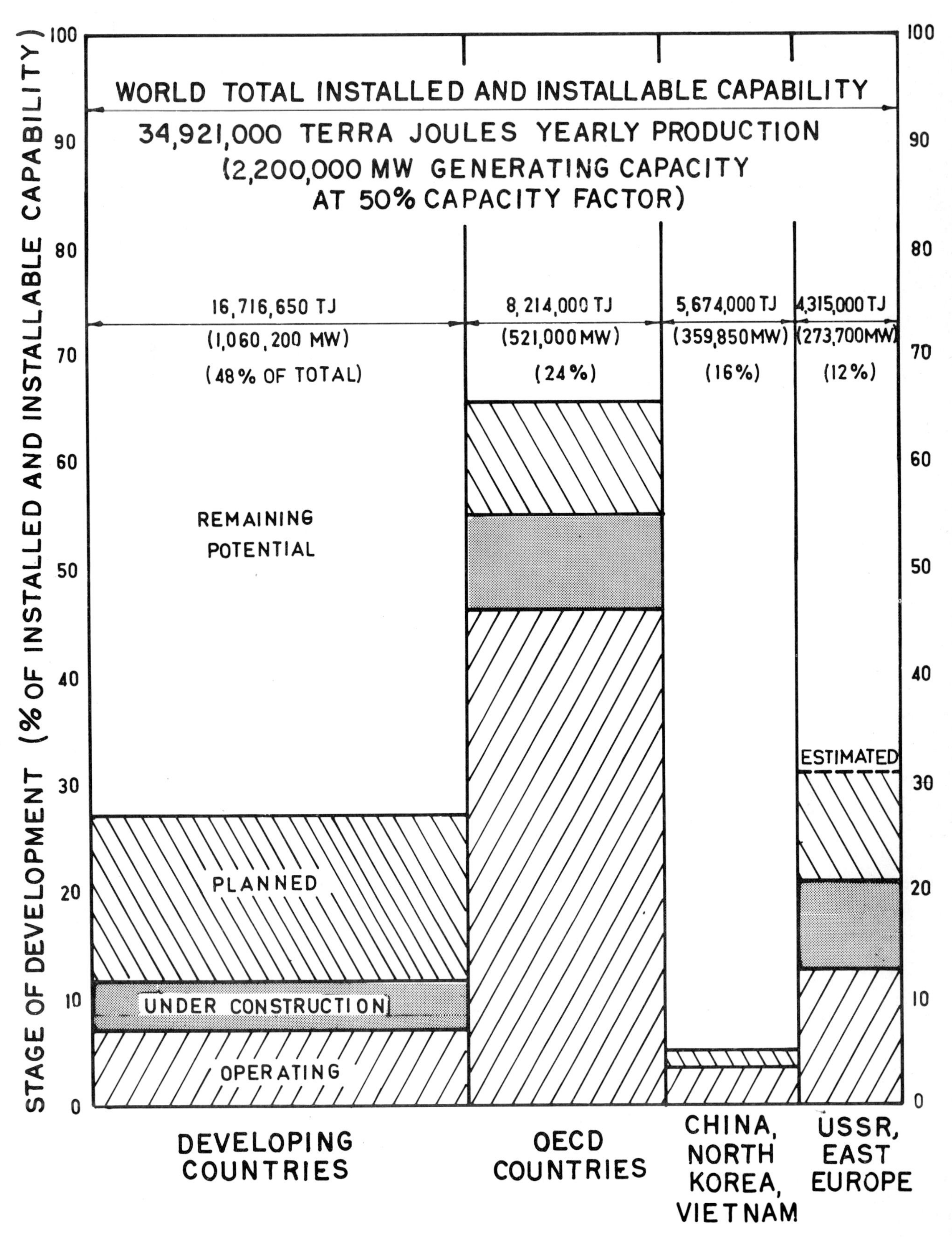
WORLD HYDRAULIC RESOURCES
STAGE OF DEVELOPMENT (% OF INSTALLED AND INSTALLABLE CAPABILITY)
WORLD TOTAL INSTALLED AND INSTALLABLE CAPABILITY
34,921,000 TERRA JOULES YEARLY PRODUCTION
(2,200,000 MW GENERATING CAPACITY
AT 50% CAPACITY FACTOR)
16,716,650 TJ
(1,060,200 MW)
(48% OF TOTAL)
8,214,000 TJ
(521,000 MW)
(24%)
5,674,000 TJ
(359,850 MW)
(16%)
4,315,000 TJ
(273,700 MW)
(12%)
REMAINING
POTENTIAL
ESTIMATED
PLANNED
UNDER CONSTRUCTION
OPERATING
DEVELOPING
COUNTRIES
OECD
COUNTRIES
CHINA,
NORTH
KOREA,
VIETNAM
USSR,
EAST
EUROPE

FIGURE 3

area that are in production, under construction, planned, and the remaining potential considered probable for development. Figure 3 shows the same data for the Organization for Economic Cooperation and Development (OECD) Countries, with 24% of the world total; the Centrally Planned Economies, 28%; and the Developing Countries, 48%. As shown, the greatest potential is in the developing countries where less than 7 percent of the probable potential is now in production, as compared to 46 percent in the OECD countries.

FACTORS AFFECTING DEVELOPMENT

There are many factors affecting the development of hydroelectric resources. A hydroelectric project generally involves the full range of water resource considerations. Usually an entire river basin is involved. Further, rarely does a project concern only a local area; most remaining potential hydroelectric development sites involve regional, country wide, and even international considerations. Irongate Dam on the Danube, built cooperatively by Romania and Yugoslavia is an example. In fact, one of the biggest problems on the multinational river basins arises from the international relationships and the international cooperation and agreements that are necessary before development can proceed. The large remaining potentials for hydropower development, such as on the Congo, the Mekong, the Uruguay, and the Parana Rivers, illustrate this problem.

Economics: Comparative cost of different sources of energy are under continual change and adjustment. Rising costs and increasing demands made on diminishing resources such as oil, natural gas and coal require a flexible approach in economic comparisons of various energy sources. The economic advantage of hydroelectric developments are increasing as the costs of fuel for heat-power generation increases, and the limitations of supply become more starkly apparent. Most methods of economic analysis tend to look backward instead of forward and thus rarely reflect the likely future.

The economic comparisons between the cost of oil and the equivalent additional, capital cost for a hydroelectric plant that would be justified over that for an oil-fired plant, are illustrated in Figure 4. In preparing the curves, the life of a hydroelectric installation was taken as 50 years, with 35 years considered to be the life of an oil-fired plant. The curves are illustrative only as they have been simplified and just show the principal cost factors. Other elements of cost, such as differences in operation and maintenance and the probable greater cost of transmission for the hydroelectric plant, are not included in the curves. These factors, however, must be considered in an analysis of a specific site.

The curves are based on the same capacity factor for the two comparison plants. While oil-fired and hydro plants in an electrical system would likely have different load curves which affect the cost of electricity, for a specific purpose the comparisons are valid. The curves show that a capacity factor of 0.53 (4650 kilowatt hours per year per kilowatt of capacity), which is about average, with an assumed cost of oil at the time of completion of the plant at 20 U.S. dollars per barrel and remaining at that cost during the 35 year life of the plant, and with an interest rate of 8 percent, then an additional capital investment of 1,650 U.S. dollars per kilowatt more than the capital cost of a kilowatt of oil-fired steam plant, could economically be made for the hydroelectric plant.

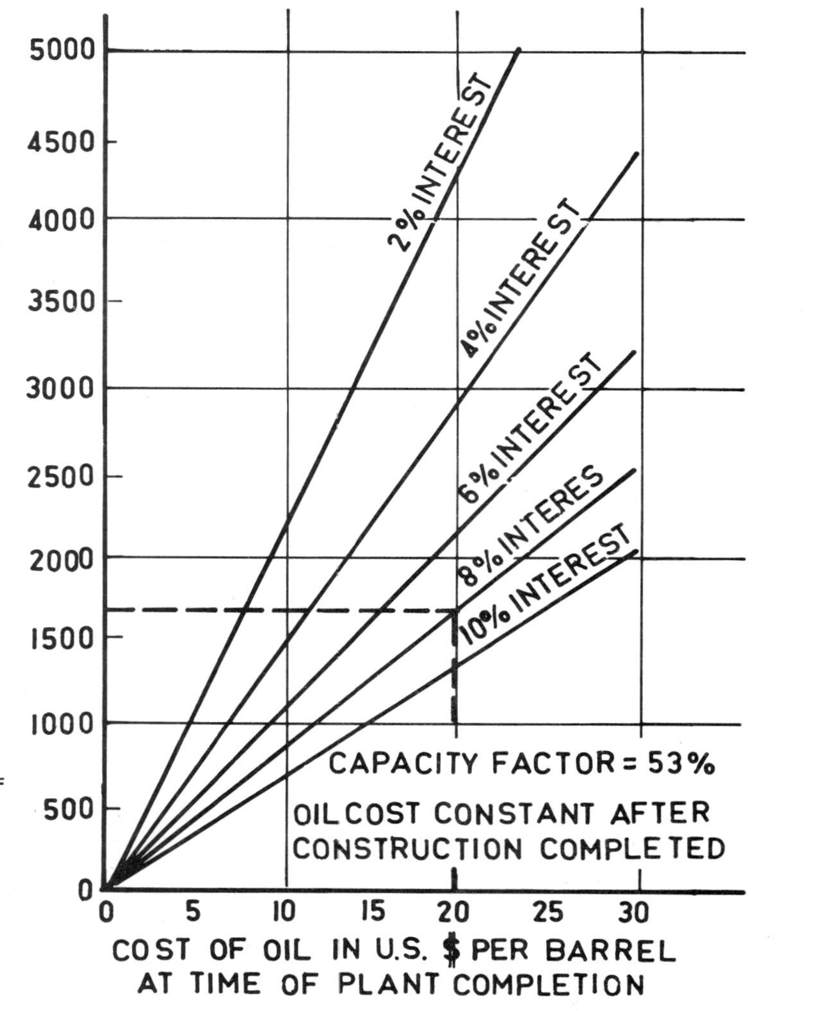
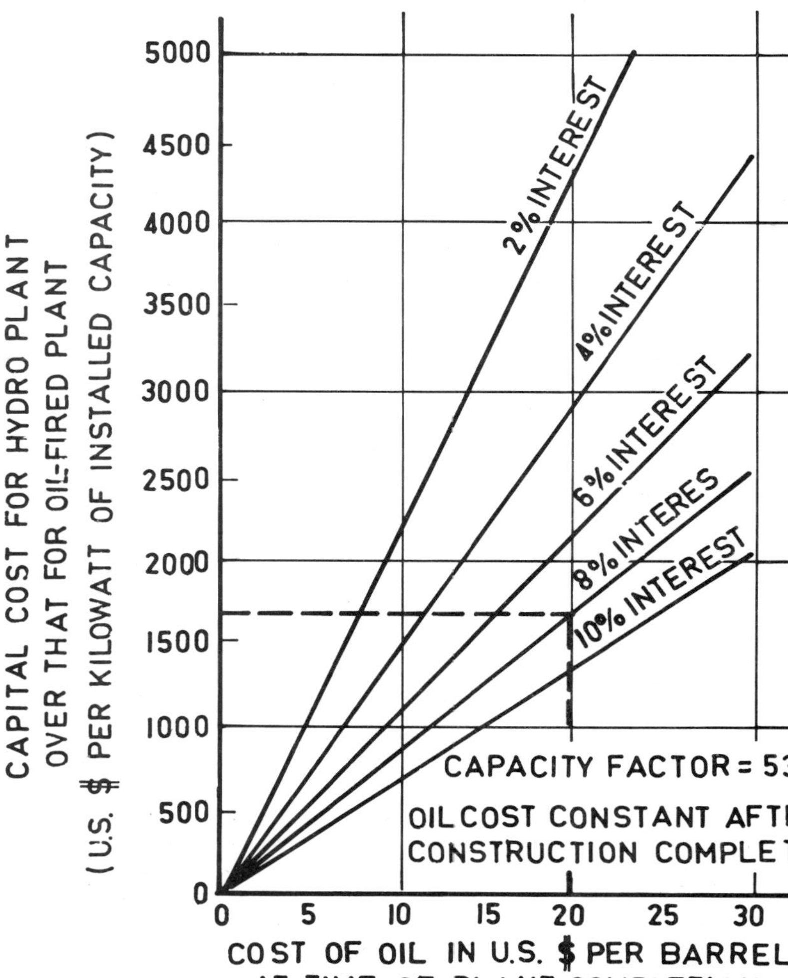

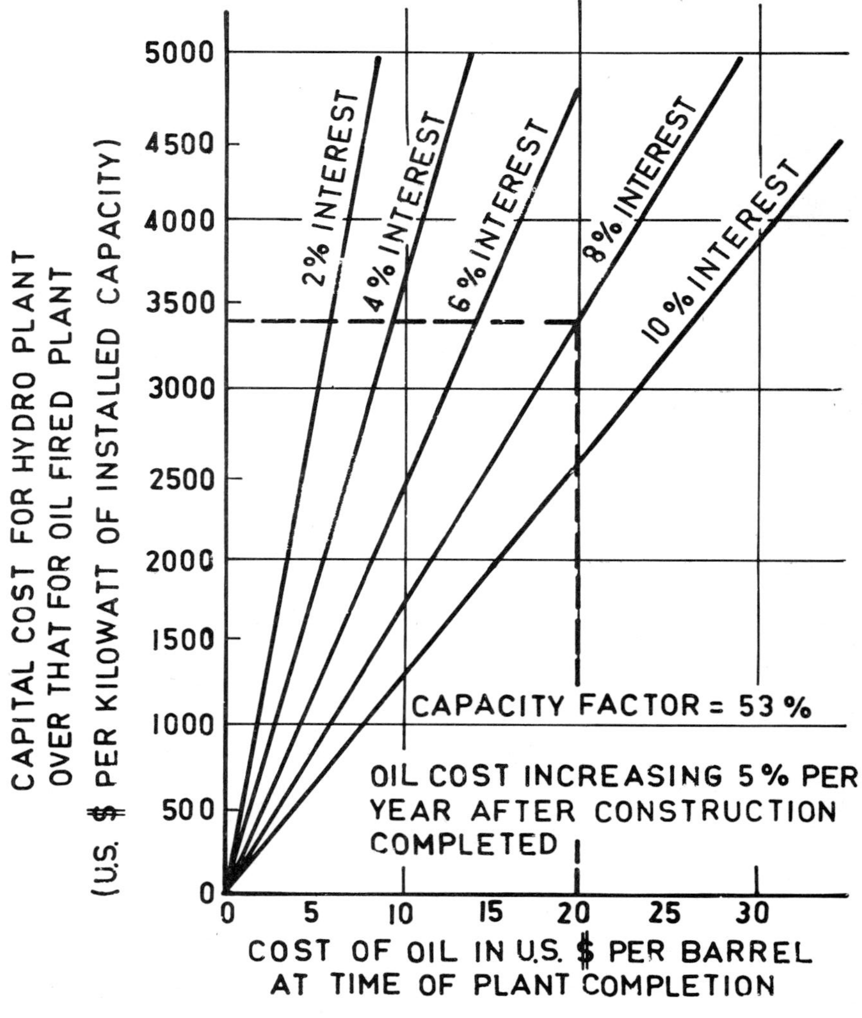

ADDITIONAL CAPITAL COST ECONOMICALLY JUSTIFIED FOR HYDROELECTRIC PLANT OVER THAT OF OIL-FIRED ELECTRIC PLANT WITH DIFFERENT OIL COSTS AND RATES OF INTEREST FOR FINANCING.

FIGURE 4

If an annual inflation rate of 5 percent is applied to the cost of the oil, again assumed at 20 U.S. dollars per barrel at the time of plant completion (and an equitable economic comparison must take this into account), then an additional capital investment of 3,400 U.S. dollars per kilowatt to that for the cost of an oil-fired steam plant could economically be made with interest costs at 8 percent. This illustrates the "inflation proof" characteristic of hydropower.

The cost of an oil-fired plant being completed today (or a coal-fired plant--the costs are presently approximately the same) is about 400 U.S. dollars to 500 U.S. dollars per kilowatt. Hydroelectric plants now considered economically viable involve costs ranging from 800 U.S. dollars to about 1500 U.S. dollars per kilowatt. These curves illustrate the need for a fresh look at the economic advantages of hydroelectric power. A recent study indicates that in the United States, a fossil fuel steam plant, starting with design today and completing construction in about eight years, has an estimated cost of 900 U.S. dollars per kilowatt of installed capacity. About one fourth of this increase is attributable to increased environmental requirements and the remainder is the expected inflation in the cost of machinery and construction.

While coal generally is a lower cost fuel than oil, the increase in costs of environmental protection and to meet safety and health requirements, is closing the gap.

There are many factors that must be included in comparing hydro and thermal power plants, and all should be considered in any specific study. However, a basic consideration is that as fuel costs increase, a greater investment can be justified for hydroelectric power. Thus more sites may now become economically feasible, as may also the upgrading or rebuilding of existing plants. A reevaluation of former studies appears well justified.

Multipurpose Use of Water: In considering any magnitude of hydroelectric development, planning must consider all water resource needs and the ways to meet the needs. Water is a basic resource that acts upon and interacts with all man's activities. Hydroelectric development cannot be undertaken in isolation from other needs.

In many areas of the world water supply is the controlling factor of present and future human activity. Thus, a comprehensive master plan for a river basin, and perhaps including that of adjacent river basins, is a must for optimum development, especially as water supply becomes a more critical consideration in such a region or area. Without such an approach, some have suggested that a water crisis of worldwide dimensions could exceed the energy crisis in threatening the future. All of man's activities require an adequate supply of water.

The value of hydroelectric energy produced in a comprehensive plan for a river basin can prove to be the economic support that makes the entire development possible, with full consideration of and facilities to meet future needs. With the basic resource of water, a plan that maximizes the immediate economic return can be a serious mistake, and may jeopardize meeting future requirements.

The assignment of costs to the various benefits of multipurpose development of water resources is complex and the methods vary widely. Returns from the hydroelectric power generated in a multipurpose project are often used to help support other desirable uses of the water, such as for irrigation. Optimizing overall benefits, within basic constraints, should be the objective. For instance, the combining of navigation requirements with power generation facilities has made successful projects of such developments as the Danube Waterway and the St. Lawrence Seaway.

Technology: As technology advances, it will affect hydroelectric and related activity in both the developing and the developed countries. Better understanding of hydrology is a necessity because of the stochastic nature of water resources. The recent advances in the art of precipitation enhancement through cloud seeding techniques, needs to be a part of the consideration in the development of a master plan for a river basin.

Today's remote control capabilities for hydroelectric plants, especially in the developed countries, makes for more efficient use of energy. Such technology also helps make the operation of small plants again economically feasible, and they can be made an effective part of either a large or a small system.

With today's high voltage transmission of electricity, a thousand miles or more of transmission is now economically and technically feasible. Voltages of over a million volts are now under consideration and may make more very large and remote hydroelectric projects feasible.

Improvements in materials for and design and fabrication of equipment can be expected. With anticipated increase in utilization of small hydroelectric generating units, improvement in design for better efficiency, and to simplify parts for mass production and to make operation simpler and easier is likely. New concepts will emerge, such as the bulb turbine and the straight flow turbine with rim type generator. Another technology advance is the slant-axis turbine-generator for medium to small size unit installations. It has the potential for substantial reduction in excavation and structural needs, and thus in overall costs of a hydroelectric plant. Other cost-saving developments are expected.

Environment: Storage and diversion structures have an effect on the regime of a stream which should be evaluated to take full advantage of positive effects and to minimize and mitigate changes that are adverse. This requires consideration of the biological system; for instance, the habitat of waterfowl is an important factor. Flood prevention is a plus; other effects such as silt and debris retention, changes in the hydraulic properties of the stream and the effects on the streambed, can have both plus and minus aspects.

The effects of a large reservoir and its operation on fishlife in a stream vary widely. On some streams, such as those with widely fluctuating flows in arid and semi-arid areas, the effects will be mostly favorable. Food production from fish, as well as recreational aspects, can be greatly increased. Blocking of the upstream migration of anadromous fish can be mitigated, at least partly, by fish ladders along the fish hatcheries and related management programs.

The aesthetics of control structures and reservoirs can be great assets. Lakes, whether natural or manmade, provide large water related recreational and other human benefits that are far reaching, especially in arid areas. They often provide opportunities for recreation to millions that otherwise would be unavailable. For instance, the recreational value of manmade reservoirs in the United States is demonstrated by the level of over 600 million visitor-days each year, several times that of visits to the nation's national parks.

Societal: The short term effects, at least on the existing social cultures in an area, will depend upon the magnitude of the water resource development. Small hydroelectric installations, in providing electricity for house and street lighting, and sometimes an assured water supply as a by-product, can initiate the change from a primitive, poverty stricken culture to improvement in living standards. A large development that may require the relocation of villages and large numbers of people present problems that have serious impacts and require full knowledge and sound judgement in determining solutions. There are no easy ways to accomplish these relocations, but often the conditions of those moved can be greatly improved with proper planning. Such problems emphasize the need for multi-discipline, full systems analysis of proposed developments, and for decisions made with mature judgements.

Legal and Political: The legal rights to water vary widely. Some areas have essentially property rights under the prior appropriation for beneficial use doctrine illustrated in irrigation areas; others have common law riparian rights, and some have little or no legal framework. As need for water approaches the limits of the capabilities of supply from the water resources, the legal problems become more difficult.

Resolving legal problems to permit development requires full information and objectivity, with well balanced concern for the overall needs of society. As in other areas, the need is for data and knowledge and then for full and proper consideration of all factors in working for solutions.

A number of international developments, such as the Columbia River and the St. Lawrence River involving Canada and the United States; the Rio Grande involving the United States and Mexico; the Irongate Dam on the Danube River involving Romania and Yugoslavia; the Parana River in South America; a number of cases in Europe, including the Rhine and Danube Rivers; the Mekong in Southeast Asia; and several others have demonstrated that arrangements can be worked out for the mutual benefit of the cooperating countries.

REQUIREMENTS FOR DEVELOPMENT

One of the difficulties facing optimum development of hydroelectric potential is that of establishing an organizational structure that can function effectively. Even the most simple projects will involve interface with governmental organizations; and most of the complex projects will be planned, constructed, and operated under government authority.

The organization must be able to function with geographic capabilities to include both generating and marketing areas of river basins, as well as the weighing of priorities from the national interest standpoint. Thus it must be a part of, or closely correlated with, the electric industry, as well as the natural resources development and management agencies.

In most of the industrial nations and some of the developing countries, organizations exist that are adequate. However, in water resource planning, more so than in any other, a big need is for people with broad based understanding of the physical (including environmental), economic, and social factors to insure balanced and mature judgements. The difficulties occur in bringing all the parts of a program into an optimum balance for the best overall interest, short and long term. This is an area which should receive greater attention from educational and training institutions, as well as from financing and governmental organizations. In developing countries, a great need also usually exists for extensive training of personnel to carry out details of planning, design, construction, and operation of projects. Adequate operation and maintenance of completed works are most important, but deficiencies often occur.

Where multi-national river basins are involved, special organizational entities are required to determine equitable sharing of costs and benefits. In developing countries, this is an area where the United Nations, or perhaps organizations such as the World Energy Conference, can be helpful in providing objective and broad-based personnel to assist with problem determinations and solutions. Financing agencies require an arrangement that will insure appropriate protection of funds and of completed facilities, and satisfactory design, construction, and management.

From a worldwide standpoint, maximum development of the hydroelectric potential is necessary, consistent with the other needs for water, including an intelligent balance with environmental requirements. Each kilowatt hour of electricity produced reduces the demand for finite energy resources. About 660 kilowatt hours of hydroelectric power at a plant almost anywhere in the world, reduces the requirement for oil by one barrel, or its fuel equivalent.

PROBABLE FUTURE DEVELOPMENT

Indications are that extensive hydroelectric development is likely to occur during the next several decades. Hydroelectric energy appears to be the most economical and feasible of the renewable sources of electric power and thus efforts to develop additional production from this source will be pushed. There are a number of constraints, some of which will have greater deterrent effect as development proceeds.

However, assuming suitable financing arrangements can be made, and this will be a problem for the developing countries, it seems reasonable to expect that the installations under construction reported in the 1976 WEC Survey of Energy Resources will be completed by 1985; and further that the planned installations will be completed by the year 2000. This would be an increase of 2-1/4 times the present worldwide capacity -- a uniform growth rate of 3.5 percent per year over the 24 year period. This overall rate divides up with the OPEC Countries growth being 1.5 percent, and the Centrally Planned Economies and the Developing Countries having a growth rate of approximately 5.6 percent.

After the year 2000, the installed and installable capacity, as reported in the WEC survey, will be limiting for the OPEC Countries, even with greater pressures to develop the maximum of renewable energy resources. There will also be greater consumptive needs for water for other purposes not compatible with maximum power generation, such as irrigation of new lands for increased food supplies. There also will likely be strong support for some free flowing streams.

It is probable that all presently feasible potentials were not included in the survey; for instance, the United States data do not include sites smaller than 5 megawatts in size -- rough estimates indicate this could be as much as 15 or 20 percent of the total. It is also probable that utilization of some sites will be precluded because of developments in the reservoir or backwater areas. In the non-OPEC countries, it is reasonable to assume development after the year 2000 would proceed at about the same rate as in the preceeding 15 years.

Taking the above factors and other considerations into account, and with full recognition of the many unpredictable variables, the following is the author's estimate of the most probable development by the years 1985, 2000, and 2020.

TABLE I

ESTIMATED PROBABLE HYDROELECTRIC DEVELOPMENT

Divisions	Potential Energy in Thousands of Ter ajoules (TJ)				
	Year 1976	Year 1985	Year 2000	Year 2020	Total Developable from 1976 WEC Survey
OECD Countries	3,776	4,493	5,369	7,800	8,214
Centrally Planned Economies	719	1,200	2,880	8,700	9,990
Developing Countries	1,172	1,973	4,490	11,800	16,717
World Total	5,667	7,666	12,739	28,300	34,921

The figures shown are the probable annual average energy from installed hydroelectric facilities for the year indicated.

Such a development, based on an estimated cost of 1,000 U.S. dollars per hydroelectric kilowatt installed, and with a capacity factor of 0.50, will require the magnitude of capital investment shown on Table II.

For the developing countries, costs shown in Table II would average about 6 percent of their estimated gross national product during the periods shown, and it is evident that financing assistance will be required from the industrial nations. Also, markets for some of the large potential in the developing countries may be slow in materializing, as they require large capital

TABLE II

ESTIMATED CAPITAL COSTS OF HYDROELECTRIC INSTALLATIONS
BASED ON TABLE I

Divisions	Average Annual Costs Between Years Shown -- 1976 U.S. Dollars in Billions			
	1976-1985	1985-2000	2000-2020	1976-2020
OECD Countries	5.05	3.70	7.71	5.80
Centrally Planned Economies	3.39	7.10	18.45	11.50
Developing Countries	5.64	10.64	23.17	15.31
World Total	14.08	21.44	49.33	32.61

investments also. However, the increasing value of hydroelectric power to the world economy will help alleviate these problems. For instance, the vast 43,000 megawatts of hydroelectric power potential on the lower Congo River in Africa is attractive to high energy-intensive industry. This could ultimately justify locating plants where the low-cost and available power can be utilized. In the first stage development, however, power is being exported over a high voltage (800 KV) transmission line for nearly 1,000 miles to a market.

In the industrial nations hydroelectric development is expected to continue as indicated, with special attention being given to expanding generating facilities at existing reservoirs and installing them at non-hydro dams. Relatively small installations have received little attention in North America in the past, but this is expected to change with the more favorable economic justification. Disagreements over the development of water resources, no doubt will continue. The difficulty is establishing priorities that will best meet the needs of people.

It is a worldwide problem and a worldwide opportunity to do a better job of utilizing natural resources. The appendix of the full report contains data on individual countries, primarily that obtained during the WEC 1976 Energy Resources Survey with some updating.

ACKNOWLEDGEMENTS

The various national committees of the World Energy Conference provided most of the basic data included in the report. The information was collected and tabulated by the WEC headquarters office in London, and published as The World Energy Conference Survey of Energy Resources - 1976. Many sources too numerous to mention were utilized in reviewing the information. The coordinating efforts of the Conservation Commission, and the valuable suggestions received from them and from the discussions at the Instabul Conference, were especially helpful. Deep appreciation is expressed to all who have contributed.

\- - - - -

THE CONTRIBUTION OF NUCLEAR POWER TO WORLD ENERGY SUPPLY, 1975–2020

J. S. Foster
Canadian National Committee, World Energy Conference

M. F. Duret, G. J. Phillips, J. I. Veeder, W. A. Wolfe
Atomic Energy of Canada Limited

R. M. Williams
Energy, Mines and Resources Canada

THE CONTRIBUTION OF NUCLEAR POWER TO WORLD ENERGY SUPPLY, 1975 TO 2020
EXECUTIVE SUMMARY

General Remarks by J.S. Foster
Vice-Chairman
Canadian National Committee
World Energy Conference
Member of Conservation Committee

Nuclear power, a term used in this report to mean power derived from the fission of nuclei of atoms of uranium and plutonium, will undoubtedly make a major contribution to man's supply of controllable energy. This contribution will depend, basically and ultimately, on the distribution of relevant natural resources and on man's relevant knowledge and industry. However, the rate at which nuclear power will become available will depend critically upon decisions respecting policies and actions pertinent to its development and application. The importance of social and political considerations and attitudes and of the consequent commerical climate has not been overlooked in performing this study. It was considered, however, that to take them into account would, apart from requiring exceptional clairvoyance in a study spanning five decades, beg the question and obfuscate the results. The study is consequently a reconnaissance of what can be, not a forecast of what will be; although we believe that mankind will be best served if, in this particular matter, what can be, will be.

The study has, for practical reasons, been limited in its scope and the results are, as a consequence, a set of representative projections rather than a complete tour d'horizon. They show the implications in terms of uranium production, isotope separation and fuel reprocessing of various courses of development of nuclear power to satisfy a particular pattern of growth in the use of nuclear power. The selected pattern of growth was derived from a projection of growth in electrical energy which was produced as part of the companion study "World Energy Demand to 2020" prepared for the Conservation Committee of the WEC by the Energy Research Group of the Cavendish Laboratory, Cambridge.

It is recognized that during the period under study nuclear power will contribute to energy supply in other sectors besides that of electrical power generation. Already, for example, it has contributed to space heating in Sweden and heavy water production in Canada. Applications involving low temperature heat can be expected to multiply and other applications involving high temperature heat will follow in due course. Nuclear power will supply energy for a growing variety of uses. Eventually even those uses which depend upon fluid hydro-carbons will be served by nuclear power as it is first applied to the extraction of natural hydro-carbons and subsequently to the synthesis of these valuable compounds from coal and conceivably even limestone. An appendix to the full report, contributed by a group at the Kernforschungsanlage Jülich GmbH and entitled "An Assessment of Nuclear Process Heat from High Temperature Reactors and the Impact on World Energy Resources", offers one perspective on this topic. Although the use of nuclear power for purposes other than the generation of electricity is important and will be increasingly so in the years after 2020, it appears that prior to that time the amount of nuclear energy involved in such uses is not likely to be as great as the range of uncertainty in the amount of nuclear energy involved in electrical power generation.

At the present time the development of nuclear power is restrained by public concern and governmental caution induced by widespread and strident challenge focussed on questions about reactor safety, environmental effects, waste disposal and the proliferation of nuclear weapons. To those with the best understanding of the subject the attendant risks are at most commensurate with those associated with other major lines of human endeavour and, indeed, are, in all probability, much less. Nuclear power plants are accumulating an outstanding safety record. Environmental effects have been minimal. Fission products, incorporated in glass, have been buried in wet, sandy soil and monitored for almost twenty years. The evidence is that if they were to remain undisturbed, this would be an adequately safe method of disposal. Can there be any serious question that the deep interment methods proposed by most nations is not a more than adequate treatment? The toxicity (through ingestion) of residual plutonium is about equal to that of the lead in the original ore and less than that of the radium. It will be more concentrated but very much better sequestered than these natural toxic materials.

Nuclear power programs have not been a route to nuclear explosions for those nations that have produced them. It is conceivable that a nuclear power program might assist a nation to attain the capability to produce nuclear explosives. It is clear, however, that the absence of a nuclear power program will not prevent its doing so. International control is clearly warranted in this field but undue interference in nuclear power programs could create the very kind of situations that it is intended to prevent. There is an acknowledged dilemma here; fissile material, in common with everything in his control, can be used by man for constructive or destructive purpose. The dilemma, however, cannot be avoided; it must be faced. This report is written in the expectation that one consequence will be universal dissemination of the benefits of nuclear power.

In preparing this report the authors had the benefit of consultation with the members of the NEA/IAEA Steering Group on Uranium Resources and received valuable comment from experts in several member countries of the World Energy Conference. This assistance was very helpful in the preparation of the report of which this is the Executive Summary.

The report has been reviewed by a very competent board established by the Conservation Commission for the purpose and by the Commission itself. Comments were also received from delegates to the 10th World Energy Conference in Istanbul, September 19 to 23, 1977. The executive summary was subsequently revised to incorporate these comments and new data received as of November 1, 1977.

THE CONTRIBUTION OF NUCLEAR POWER TO WORLD ENERGY SUPPLY, 1975 TO 2020 - EXECUTIVE SUMMARY

M.F. Duret, G.J. Phillips, J.I. Veeder, W.A. Wolfe
Atomic Energy of Canada Limited

R.M. Williams
Energy, Mines and Resources Canada

Nuclear power can make a significant contribution to world energy supply. At present this contribution is most conveniently supplied in the form of electrical energy. However, in the future there will be an increasing incentive to supply some fraction of the energy whose end use is low grade or high grade heat by nuclear means. One area of possible exploitation is, for example, the manufacture of synthetic fluid fuels. The potential magnitude of the substitution of synthetic fuels for petroleum and natural gas has been estimated by a group at Jülich and these estimates suggest that by the year 2020 about 10 billion tons of oil and natural gas could be saved by using high temperature gas cooled reactors to convert coal to fluid hydro carbons. This would entail the consumption of about half that much coal and about a quarter of a million megagrams of uranium. This is a small fraction of the needs predicted for electricity generation, so numerical results in this summary are based on demand for electrical energy.

Growth of Electrical Power

The estimates of nuclear industrial growth presented here are based on preliminary assessments of electricity demand prepared for the WEC Conservation Commission as part of its overall study on world energy supply to 2020[1]. Although the demand assessment was prepared for eleven world regions as prescribed by the Commission for its study (see Figure I-1), it is subdivided here for conciseness, into three regional groups of countries, as shown in Table I-1.

[1]Eden, R.J., Private Communication.

FIGURE I-1

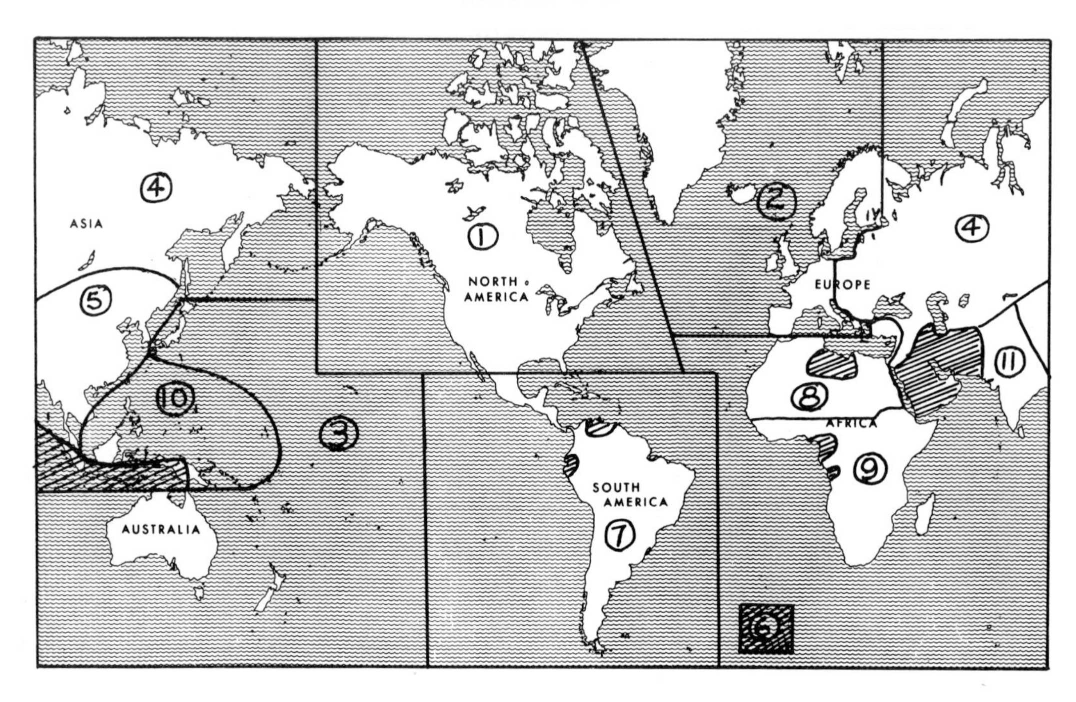

Table I-1
ESTIMATED POTENTIAL WORLD ELECTRICAL DEMAND
(10^{18} Joules Electrical Output)

	World Regional Groupings	1972	1985	2000	2020	Average Growth Rate 1972-2020
1.	OECD (Regions 1-3)	14.1	24.6	48.4	108.4	4.2%
2.	Centrally Planned Economies (Regions 4-5)	4.7	11.1	38.6	83.7	6.0%
3.	Remainder (Regions 6-11)	1.63	4.1	11.6	45.6	6.9%
	TOTAL	20.5	39.8	98.6	237.7	5.1%

Table I-2
PROJECTED WORLD NUCLEAR POWER INSTALLATION
(GW(e))

	World Regional Groupings	1975	1985	2000	2020
1.	OECD (Regions 1-3)	68	247	955	2423
2.	Centrally Planned Economies (Regions 4-5)	7	33	402	1610
3.	Remainder (Regions 6-11)	1	23	186	1000
	TOTAL	76	303	1543	5033

Growth of Nuclear Power

Nuclear power now provides about 4% of the world's electrical requirements, mostly in the industrially advanced countries. To estimate how rapidly nuclear technology could replace conventional technology for the production of electricity, a simple mathematical model suggested by Fisher and Pry[2] to describe industrial change was adopted in a modified form. The nuclear reactor growth rate in 1975 was selected to agree with the historical rate and, where known, with the reactor construction program in each region. For the long term it was assumed that the fraction of electrical energy supplied by nuclear power would tend asymptotically to 50% in all regions. This assumption is obviously not defensible on a country-by-country basis. As an average pertaining to many countries, we feel it is conservative, since considerable growth in the non-nuclear sector will be required to maintain a 50% share in the production of electricity until the year 2020. Results from this model are summarized in Table I-2.

In the near term (before 2000), these estimates lie close to the lower limit of other projections which have been made recently[3,4,5]. Beyond the year 2000, they tend to lie midway between other recent "high" and "low" forecasts[3].

Nuclear Scenarios

Nuclear power is viable and commercially competitive now. In 1975 the world's installed capacity was 76 GWe of which 80% was provided by light water reactors, 12% by gas-cooled graphite-moderated reactors and 4% by heavy water reactors, all of which operate on the once-through uranium cycle, where ^{235}U is the principal fissioning isotope. The remaining 4% of the capacity was provided by prototypes of the liquid metal cooled fast breeder reactor (LMFBR) and by high-temperature gas-cooled reactors (HTGR).

Obviously nuclear power will grow in different ways in different countries but most of the growth indicated in Table I-2 will come from technologies which are already known. The following five scenarios attempt to cover a wide range of nuclear growth possibilities.

---

[2] Fisher, J.C., Pry, R.H., A Simple Substitution Model of Technological Change, Industrial Applications of Technological Forecasting, John Wiley, 1971.

[3] Haussermann, W. et al, Demand and Supply Estimates for the Nuclear Fuel Cycle, IAEA-CN-36/493.

[4] Lane, J.A., et al, Nuclear Power in Developing Countries, IAEA-CN-36/500.

[5] Lantzke, U., World Energy Supply and demand and the Future of Nuclear Power, IAEA-CN-36/583.

References 3-5 were presented at the Salzburg Conference on Nuclear Power and its Fuel Cycle, May 2-13, 1977.

A. The "no reprocessing" scenario - in this case only thermal converter reactors (TCR) are installed in all regions. Light water reactor characteristics are used for these converters, since this is the most prevalent commercial reactor-type in use today.

B. Delayed breeder introduction - this is the same as base case C defined below, except that introduction of the breeder is delayed ten years in each region.

C. Base case - commercial breeders with a fuel doubling time of 24 years (at 100% load factor) are installed in 1993 in North America, in 1987 in Western Europe, in 2000 in Japan, and in 1995 in the USSR.

D. Improved breeder scenario - this is the same as case C, except that the breeders are assumed to have a doubling time of 10 years (at 100% load factor).

F. Thorium cycle scenario - in this case the thorium cycle is introduced instead of fast reactors. The timing and extent of the introduction are the same as in scenario C, the base case. Heavy Water reactor (HWR) fuel cycle parameters are used. The remainder of the nuclear system consists of LWR's.

Uranium Demand

Requirements for uranium in the short-term (2000) and in the long-term (2020) are shown for each of the nuclear growth scenarios in Tables I-3 and I-4 and in Figure I-2*. Only about a decade is required until annual uranium demand levels off in the OECD countries if the breeder is introduced there according to the schedule in scenario C. Development of advanced fuel cycles such as the thorium cycle and the fast breeder is being pursued vigorously in some of the developing countries, for example, India. However, it was assumed that commercial exploitation of fast breeders would occur first and to the greatest extent in the industrially advanced countries, with most of the developing countries relying on the well established converter technology. Thus, in scenario C, the accelerating uranium requirements of the developing countries reduces the impact of breeder introduction on world requirements which continue to increase although at a lower rate. Total uranium to operate the then installed nuclear plants for the remainder of their 30-year life, are shown in Table I-5. In scenario F, requirements for thorium are much lower than for uranium, rising to some 12,000 Mg (tonnes) thorium per year by the year 2020. Uranium requirements for the thorium cycle lie between those for the "TCR only" scenario and the "TCR and FBR" scenario.

---

*Regions 4 and 5 are omitted from all illustrations to facilitate comparison between uranium supply and requirements, since supply data are not available for those regions.

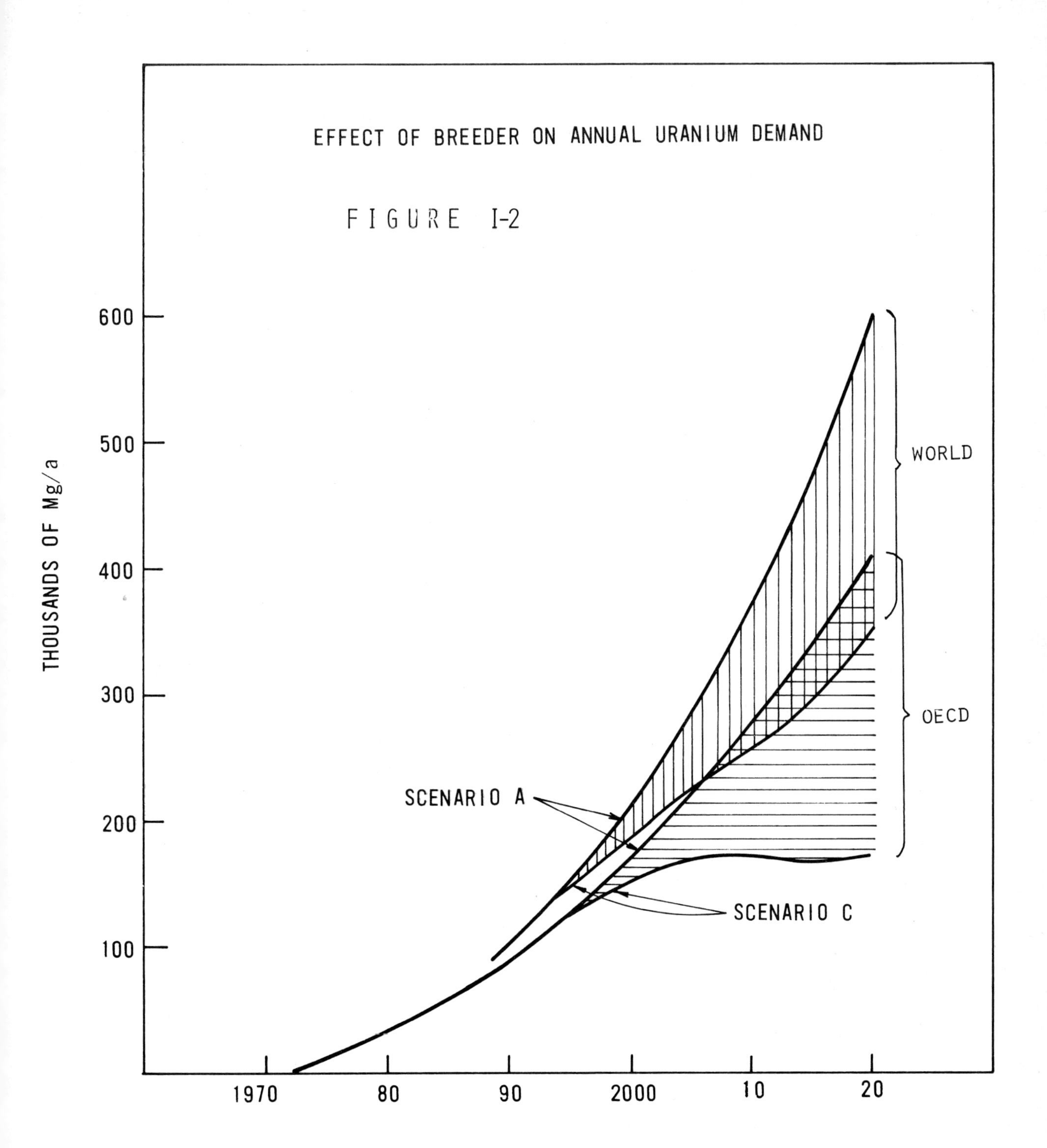
EFFECT OF BREEDER ON ANNUAL URANIUM DEMAND
FIGURE I-2
THOUSANDS OF Mg/a
600
500
400
300
200
100
WORLD
OECD
SCENARIO A
SCENARIO C
1970
80
90
2000
10
20

Table I-3
ESTIMATED ANNUAL WORLD URANIUM DEMAND
[10^5 Megagrams (tonnes)]

Nuclear Growth Scenario	OECD (Regions 1-3)		Centrally Planned (Regions 4-5)		Remainder (Regions 6-11)	
	2000	2020	2000	2020	2000	2020
A	1.8	4.2	.8	2.8	.4	1.8
B	1.7	2.5	.7	2.0	.4	1.8
C	1.5	1.7	.7	1.7	.4	1.8
D	1.4	0.8	.7	.9	.4	1.8
F	1.5	2.6	.8	3.1	-	-

Table I-4
ESTIMATED CUMULATIVE WORLD URANIUM DEMAND*
[10^6 Megagrams (tonnes)]

Nuclear Growth Scenario	OECD (Regions 1-3)		Centrally Planned (Regions 4-5)		Remainder (Regions 6-11)	
	2000	2020	2000	2020	2000	2020
A	2.0	7.8	.6	3.9	.5	2.2
B	2.0	6.6	.6	3.6	.5	2.2
C	2.0	5.3	.6	2.8	.5	2.2
D	2.0	4.6	.6	2.3	.5	2.2
F	2.0	5.85	.6	3.2	-	-

*Cumulative from 1975 to year indicated.

Table I-5
ESTIMATED WORLD URANIUM COMMITMENTS TO 2020
[10^6 Megagrams (tonnes)]

Nuclear Growth Scenario	OECD (Regions 1-3)	Centrally Planned (Regions 4-5)	Remainder (Regions 6-11)
A	13	7.7	4.8
B	9	6.6	4.8
C	7	4.7	4.8
D	5.3	3.2	4.8
F	9	5.8	-

Uranium Resource Estimates

The history of the uranium industry is relatively short and erratic compared to that of coal and hydrocarbons, aggregating less than 25 years. The industry's initial spectacular growth in the 1950's in response to an apparent unlimited demand for defense purposes, was followed abruptly by a period of rapid decline. The promise of a demand for nuclear power purposes was slow to materialize and the industry suffered for several years due to surplus inventories and excess capacity. Not until after the oil crisis of 1973-74 could the industry look forward to a more tangible future and regain the upward momentum it had achieved in the late 1950's. This short but eventful history has been responsible not only for moulding the present structure of the industry but for contributing, perhaps more than any other factor, to our present limited knowledge of world uranium resources.

Efforts to prepare global estimates of uranium resources are hampered by a number of additional factors, ranging from fundamental problems of resource classification to the simple lack of information and acceptable methodology with which to generate the estimates. The most comprehensive program to assess the world uranium supply situation on a regular basis is that carried out by the Nuclear Energy Agency (NEA) of OECD, jointly with the International Atomic Energy Agency (IAEA). Current estimates of recoverable world* resources of uranium, based largely on the NEA/IAEA's most recent (1977) survey, are summarized in Figure I-3.

The resources are divided into two classes of reliability, based on their assurance of existence, and estimates are presented of tonnages judged to be recoverable at costs up to $130/kg U ($50/lb U_3O_8).** No attempt has been made to quantify world resources exploitable at costs in excess of $130/kg U, since only for isolated cases do such estimates exist. Resources in the Estimated Additional category refer only to those which are expected to occur in areas that are relatively well known and which are associated with known deposits.

Over 60% of the world's Reasonably Assured Resources occur in North America and Africa, south of the Sahara largely in sandstone-type deposits, quartz-pebble conglomerates, and veins and related deposits. It is pertinent to note that some 77% of the total shown for Western Europe is contained in Swedish alum shales, the future exploitability of which is expected to be limited. The bulk of the remaining Reasonably Assured Resources occur in Australia, largely in vein and related deposits.

---

*Excluding the USSR and Eastern Europe (Region 4), and China et al (Region 5); Yugoslavia is included in Region 2; OPEC countries (Region 6) are redistributed on a geographical basis. All illustrations use "world" to mean "world excluding regions 4 and 5".

**Quantities of resources are expressed in metric tons (tonnes) of uranium metal (U); 1 tonne U = 1.2999 short tons of uranium oxide (U_3O_8); the unit used for international uranium commerce continues to be a pound of U_3O_8, thus $1/lb U_3O_8 = $2.6/kg U; 1 tonne = 1 megagram (Mg) U, as used elsewhere in this report, with respect to "back-end" aspects of the fuel cycle. Resource estimates are as of January 1977 with costs expressed in 1976 dollars.

FIGURE I-3
ESTIMATED WORLD RESOURCES OF URANIUM RECOVERABLE AT COSTS UP TO $130/KG U AS OF JANUARY 1977

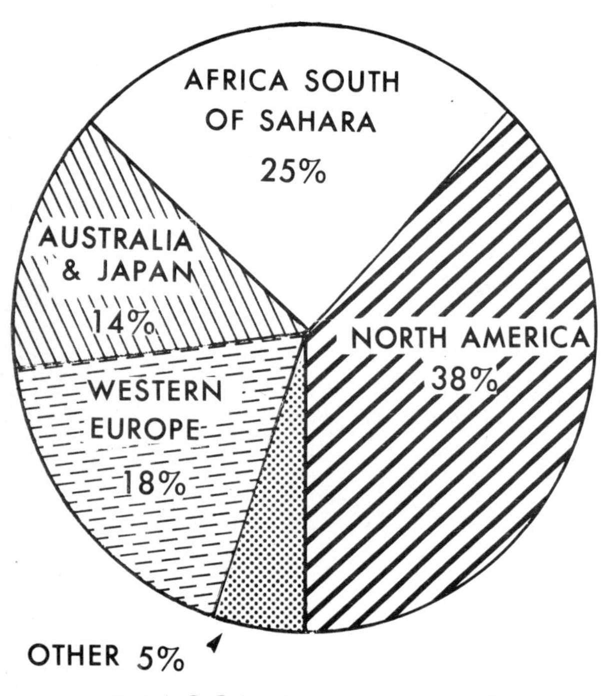

REASONABLY ASSURED RESOURCES

NORTH AMERICA 79%

OTHER 21%

ESTIMATED ADDITIONAL RESOURCES

(tonnes uranium)	World Region	(tonnes uranium)
825,000	1. North America	1,709,000
389,300	2. Western Europe	95,400
303,700	3. Australia, N.Z. & Japan	49,000
64,800	7. Latin America	66,200
32,100	8. Middle East & N. Africa	69,600
544,000	9. Africa S. of Sahara	162,900
3,000	10. East Asia	400
29,800	11. South Asia	23,700
2,191,700	Total World	2,176,200

FIGURE I-4

WORLD URANIUM PRODUCTION TO 1976

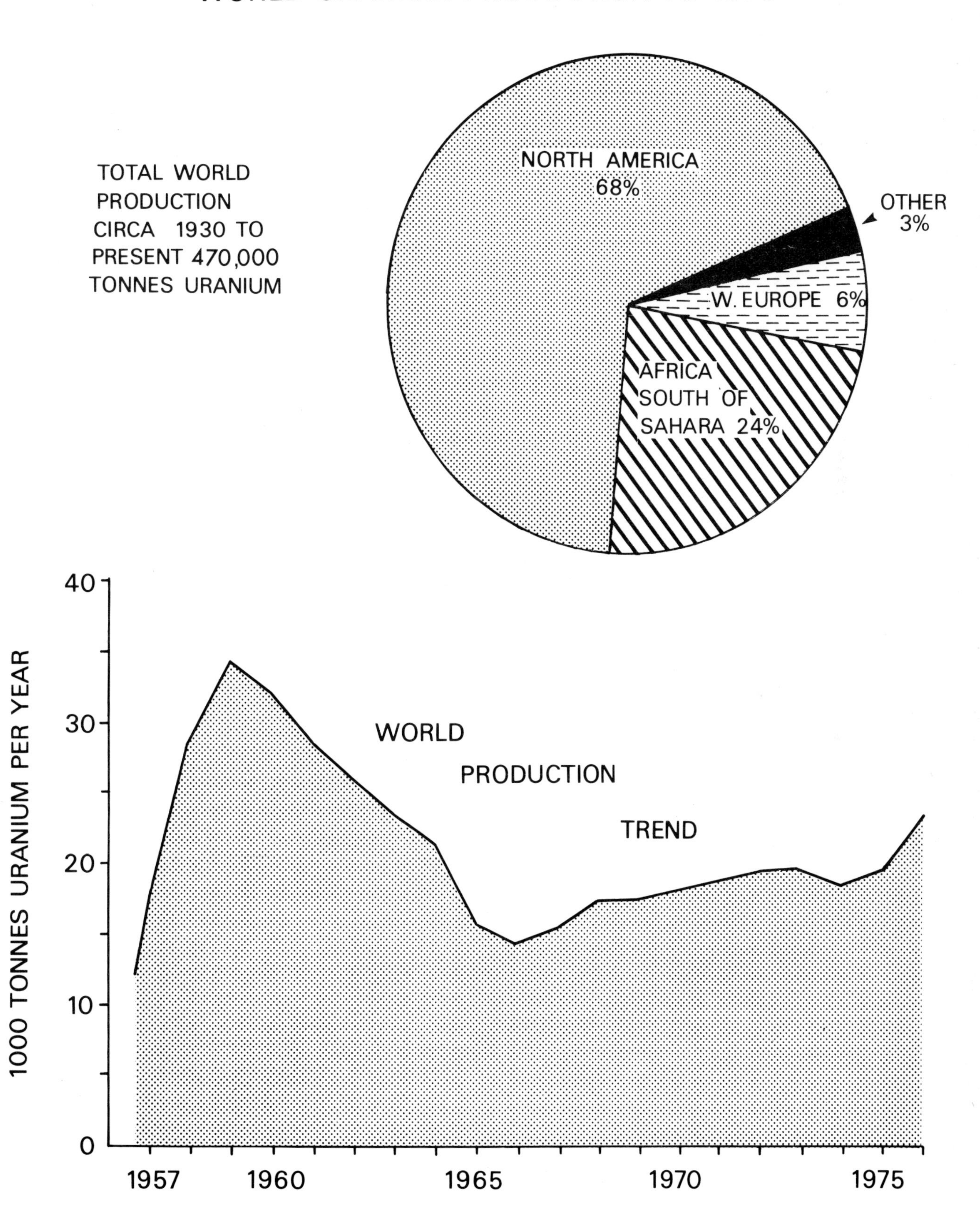

Some 79% of the world's Estimated Additional Resources at up to \$130/kg U are located in North America, slightly more than half of which are in the United States. The bulk of the world total is contained in the same three deposit-types mentioned above. The disproportionate size of the estimates for North America is in large measure a reflection of the amount of effort in terms of exploratory expenditures that has been spent in this region, compared with other regions of the world.

Finally, it is important to observe that over 70% of the resources in both categories of reliability are believed to be recoverable at cost up to \$78/kg U (\$30/lb U_3O_8) and thus can be viewed as of economic interest today. While the uneven distribution of the resources between the two cost categories is due partly to the geological nature of known resources, it can also be attributed partly to the lack of available data on resources in the higher cost category.

Uranium Availability

In order to place these estimates of known uranium resources in proper perspective, it is necessary to have some idea of the levels of production that they are capable of supporting. The world industry has, to 1976, produced some 470,000 tonnes of uranium, some 68% of which has come from North America. The bulk of the remaining production has come from three countries in Africa, south of the Sahara (see Figure I-4). Present production is some 23,000 tonnes uranium a year, compared with the industry's peak year in 1959 when production totalled some 34,000 tonnes.

Present production capability is estimated at some 33,300 tonnes of uranium a year, again largely distributed between North America and Africa, south of the Sahara. Certain production facilities are undergoing expansion and certain new plants are either under construction or planned, which if completed could bring world capability to some 55,100 tonnes uranium a year by about 1980. Beyond 1980 known resources are believed capable of supporting a maximum level of production approaching 110,000 tonnes uranium, achievable by 1990 (see Table I-6). Levels of production much in excess of 110,000 tonnes uranium a year are unlikely to be attained, without the identification of new sources of production.

FIGURE I-5

ESTIMATED WORLD URANIUM PRODUCTION CAPABILITY TO 1990

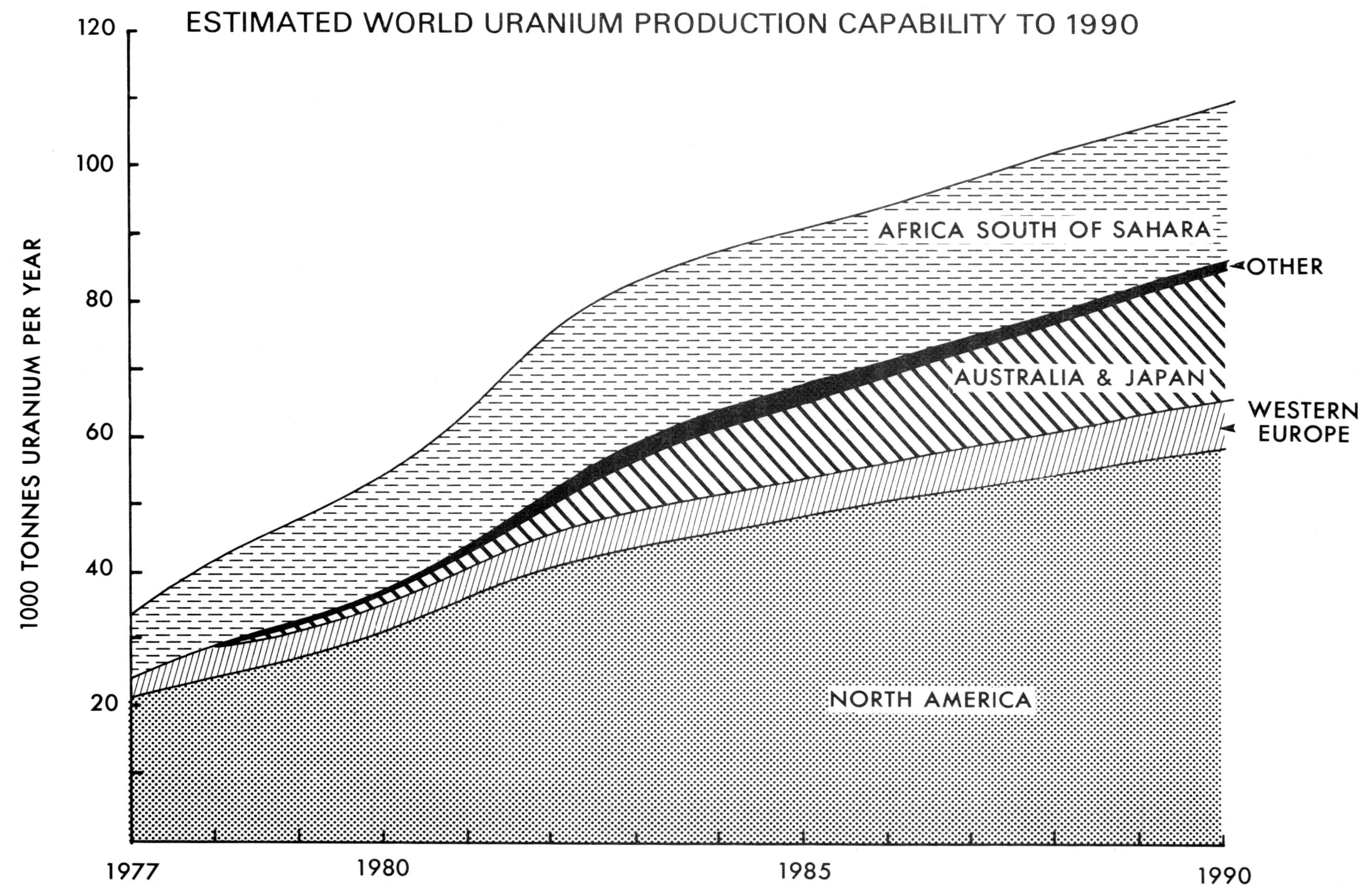

Table I-6
ESTIMATED WORLD URANIUM PRODUCTION CAPABILITY TO 1990
(tonnes uranium per year)

	World Region	1977	1980	1985	1990
1.	North America	20,800	31,550	48,500	58,250
2.	Western Europe	2,560	3,850	5,940	6,700
3.	Australia & Japan	430	530	11,030	20,030
7.	Latin America	230	1,070	1,990	600
8.	Middle East & North Africa	-	100	100	100
9.	Africa south of Sahara	9,310	18,000	23,700	23,200
10.	East Asia	-	30	..	..
11.	South Asia	..	..	..	..
	Total World	33,330	55,130	91,260	108,880

\- nil
.. no estimate available

These projections of production capability, however, are subject to a number of constraints. Certain physical limitations on the rate that particular deposits can be exploited, as well as the realities of declining grades and resource depletion, have been incorporated into the projections. Constraints have also been taken into account in a few cases where production rates in certain non-exporting countries are set by government policy and where there are very clear limitations to availability (e.g. Swedish shales). Several other types of contraints, however, are much more elusive, not easily quantifiable, and may possibly have a negative effect on the projections. Assumptions made about the timing of uranium developments in Australia*, for example, may prove to be optimistic, supplies of manpower and equipment may prove to be inadequate, increasing delays may be experienced due to more complex licensing and regulatory requirements, appropriate forward base-load sales contracts may be elusive, and adequate financing may not be forthcoming. The projections illustrated in Figure I-5, therefore, must be viewed as the maximum level of production that known resources are capable of supporting.

Unconventional Uranium Resources

The bulk of present knowledge of the world's uranium resources is associated with the so-called "low-cost" resources - i.e. recoverable at up to \$78/kg U (\$30/lb U_3O_8) - largely since it is these types of resources that have to date been exploration objectives. Resources recoverable at costs that lie in the range of \$130/kg U (\$50/lb U_3O_8) to well over \$260/kg U

*All uranium development in Australia is currently held up pending implementation of new government policy, which will take into account the findings (released in June 1977) of the Ranger Uranium Environmental Inquiry.

($100/lb U_3O_8) have also received some attention, partly for academic reasons and partly because they represent fall-back alternatives, should exploration for conventional sources be unsuccessful. In addition to these very-high cost sources of uranium, there are several types of operations from which uranium can be recovered as a by-product, some of which have recently begun to provide small but appreciable quantities to the uranium supply stream.

Uranium can be recovered, for example, as a by-product of phosphoric acid production, from solutions generated in the leaching of certain copper ores, from monazite which is produced in the heavy-mineral beach-sand industry, and as a co-product of several other elements from Swedish alum shales. The possibilities of recovering uranium as a principal product from phosphate rock, marine black shales (e.g. Chattanooga shales), above-average grade granites, coals and lignites, and sea-water have also been investigated.

Although many of these "unconventional" sources are very large in terms of a resource, their potential for contributing large increments of annual production is in most cases limited. The constraints vary from case to case and include such things as the lack of technology for recovery, the high cost of production even when technology is known, the dependence on the rate of production of the principal or co-product, the often vast scale of mining required for the very low-grade sources, and the environmental implications associated with exploitation. In view of these various and complex factors, it is expected that such unconventional resources will be capable of contributing only incremental amounts to world uranium supply. Except in the case of uranium produced as a by-product or co-product, costs are expected to exceed $130 and in most cases $260 per kg U ($50 and $100/lb U_3O_8). Production units will be small in terms of uranium output, probably in the range of 100 to 300 tonnes U/year.

Undiscovered Uranium Resources

Comprehensive quantitative estimates of undiscovered uranium resources, which may occur beyond those in the NEA/IAEA's category of Estimated Additional Resources, have not been made on a global scale. Indeed, the United States is the only country which has published such estimates, the most recent being some 1.5 million tonnes uranium, in addition to tonnages classified as Reasonably Assured and Estimated Additional Resources. In the case of most countries, the need to provide comprehensive assessments of their undiscovered uranium resources has only recently become evident, and efforts to do so have been severely hampered by both the lack of basic geological, geochemical and geophysical information, and the embryonic state of the art of methodology for estimating undiscovered, recoverable resources.

In the absence of a comprehensive data base, several attempts have been made to estimate the world's undiscovered, recoverable resources using mathematical/statistical models. Various estimates have been reported ranging from 80 to 280 million tonnes of uranium. Although the results of these efforts are encouraging, they are in no way conclusive. Indeed, methodologies employing mathematical/statistical approaches, particularly those based on the "life-cycle of production" and "discovery rates", continue to be viewed with skepticism by most uranium resource experts. By an examination of known uranium resources in relation to the world's geologic provinces, however, one can readily conclude that it is not logical to assume that North America, with only 17% of the world's* land mass, should prove to have 58% of the total estimated known resources (Reasonably Assured plus Estimated Additional Resources).

Although uranium exploration activity has been reported in some 80 countries, only in the case of a small number of these could it truly be said that the activity was substantial. Indeed, except perhaps for North America and Western Europe, no region of the world has received nearly the attention its potential deserves. The overall problem is not likely to be one of existence of uranium resources, but one of availability on the required time scale. Access to many of the world's favourable areas is restricted by lack of infrastructure and often by regulations which do not permit foreign participation in uranium exploration. Moreover, present and even developing uranium exploration technology may prove inadequate for the task of discovering uranium deposits that are expected to occur at depth.

Measures of Required Effort

Considerable effort will be required to make uranium available from known sources at the rates illustrated in Figure I-5. Although the resources needed to support this capacity have largely been discovered, further effort will be required to identify and delineate resources in the Estimated Additional category. It is also essential to note that financial commitments have yet to be made for the projected additional capacity beyond about 1980. Moreover, the great bulk of production required to meet the growing production gap, which will likely commence in the late 1980's, will need to come from new sources.

There are a number of ways to illustrate the amount of effort that will be needed to make the required uranium available, including measures of energy, equipment, materials and manpower. Price is a measure that incorporates all of these ingredients, the most important elements being cost of discovery, development, plant construction, and production, as well as those costs associated with the use of money (i.e. rate of return on investment). Present world prices are in

*Excluding the USSR and Eastern Europe (Region 4), and China et al (Region 5).

FIGURE I-6
SCHEMATIC ILLUSTRATION OF WORLD URANIUM SUPPLY PROBLEM, 1980 TO 2020

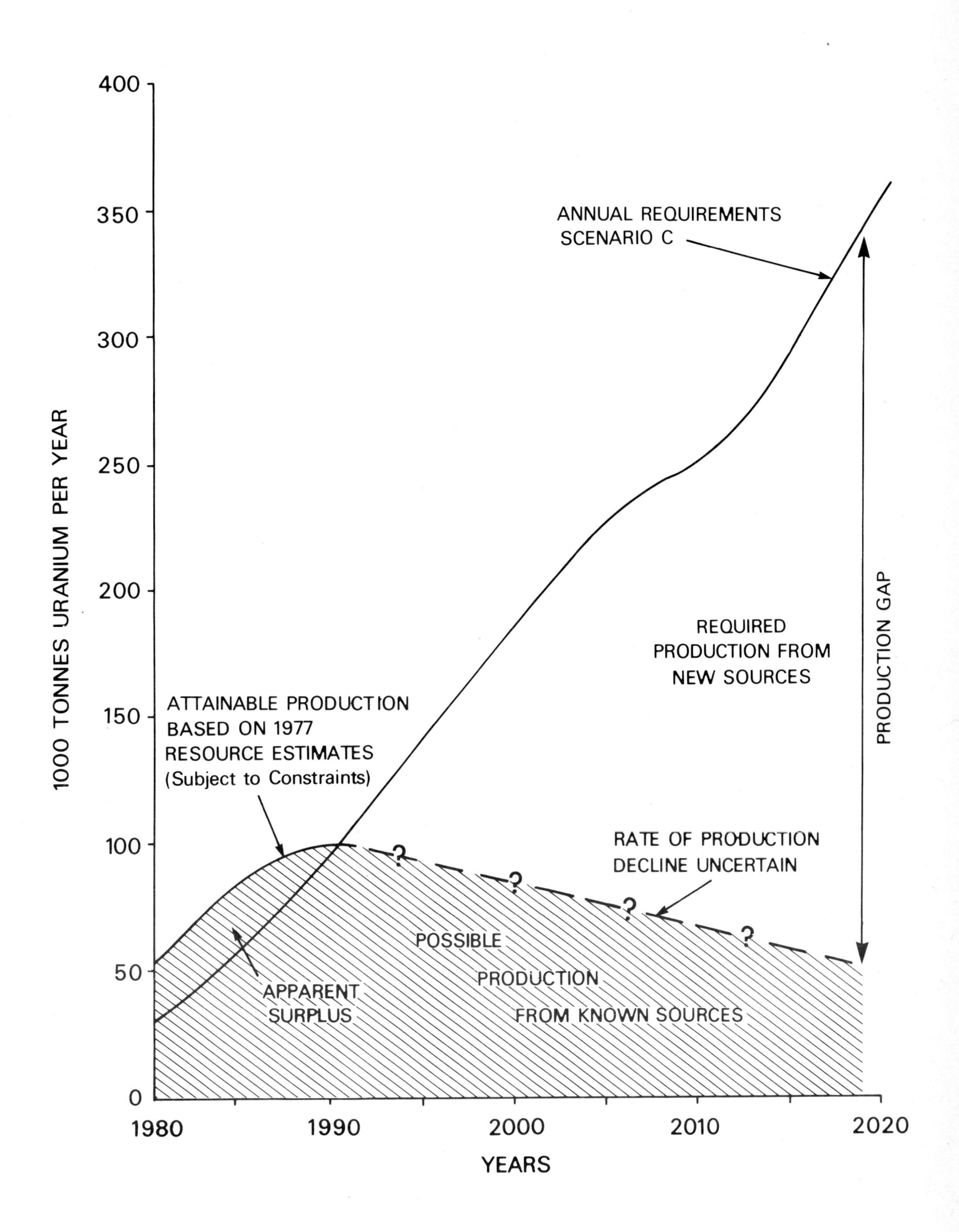

FIGURE I-7

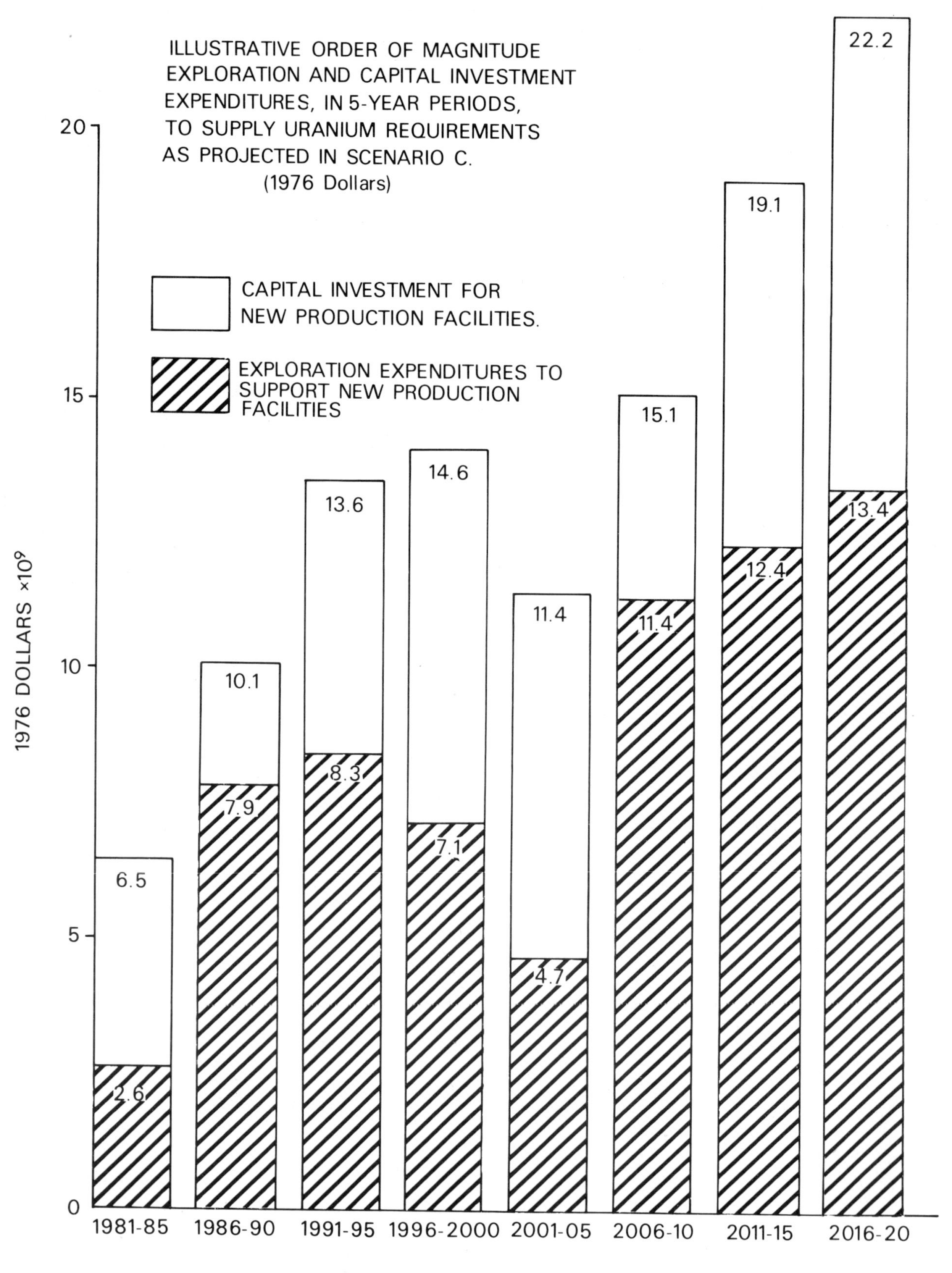

ILLUSTRATIVE ORDER OF MAGNITUDE
EXPLORATION AND CAPITAL INVESTMENT
EXPENDITURES, IN 5-YEAR PERIODS,
TO SUPPLY URANIUM REQUIREMENTS
AS PROJECTED IN SCENARIO C.
(1976 Dollars)
CAPITAL INVESTMENT FOR
NEW PRODUCTION FACILITIES.
EXPLORATION EXPENDITURES TO
SUPPORT NEW PRODUCTION
FACILITIES
1976 DOLLARS ×10⁹
20
15
10
5
0
6.5
2.6
10.1
7.9
13.6
8.3
14.6
7.1
11.4
4.7
15.1
11.4
19.1
12.4
22.2
13.4
1981-85
1986-90
1991-95
1996-2000
2001-05
2006-10
2011-15
2016-20
FIVE-YEAR PERIODS

the order of \$104/kg U (\$40/lb U_3O_8), a level which this study believes is sufficient (in 1976 dollars) to provide the required incentive to permit the industry to meet future requirements through the discovery of additional conventional-type uranium resources*. Clearly, there can be any number of price projections, depending on the assumptions used and the message the estimator wishes to convey.

Price projections, however, do not successfully place in perspective the great magnitude of effort that will be required to make the required uranium available. Indeed, such projections can lead to a sense of false complacency by inferring that everything is possible given a price. Such a conclusion is particularly dangerous in view of the theoretical fact that the energy content of uranium is good value even at costs of \$260/kg U (\$100/lb U_3O_8).

The nature of the world supply challenge is illustrated schematically in Figure I-6, in the context of projected annual world uranium requirements, using assumptions described for Scenario C. Production rates from known sources will decline after about 1990, due to depletion of resources in some deposits and to the mining of lower grade tonnages in others. Even assuming that the estimated maximum capability of about 100,000 tonnes uranium a year from known sources can be achieved, some 300,000 tonnes of annual production would have to be developed from new sources between 1990 and 2020.

Although a portion of the production gap illustrated in Figure I-6 will be met from currently-known subeconomic sources, by lowering costs and through advances in technology, the bulk of the additional requirement must be met by production from new sources**. The actual quantity of uranium resources that must be discovered will, however, vary considerably depending on the order in which deposits of different sizes and grades will be discovered and developed. By using as "measuring sticks" the resources and production capabilities of a number of known deposits, the order of magnitude of discoveries needed to support the required new production can be put into better perspective.

The results of such an exercise are summarized in Table I-7. Six mine-types of varying sizes and grades (both open-pit and underground) were chosen as illustrative examples of the kind of discoveries that could be forthcoming to provide the required additional production, which is schematically illustrated in Figure I-6. Clearly, quite a different mixture of mine-types could have been chosen and both the order and

---

*Prices are expected to rise modestly, however, as exploration objectives become deeper and as exploration efforts concentrate in more remote areas.

**Extensions of known deposits will normally extend the life of existing operations rather than increase their production capability. Incremental additions to supply may also be met from unconventional resources, as noted earlier.

number of discoveries could have been varied; the object was simply to present an order-of-magnitude illustration in terms that can be readily identified with known producing operations.

Table I-7
ILLUSTRATIVE DISCOVERY REQUIREMENTS TO MEET ANNUAL WORLD URANIUM DEMAND TO 2020 UNDER SCENARIO C

	If all additional uranium production to the year 2020 is to be met from the finding and subsequent opening of a mix of new mines similar to:	Number of Discoveries Required by the year 2015
A	Large-sized, low-grade, open-pit mine; e.g. Rossing, Namibia	12
B	Medium-sized, medium-grade, sandstone production centre; e.g. New Mexico, USA	78
C	Large-sized, low-grade, underground mine; e.g. Denison, Canada	8
D	Medium-sized, medium-grade, underground mine; e.g. Eldorado, Canada	60
E	Large-sized, high-grade, open-pit mine; e.g. Ranger, Australia	27
F	Small-sized, medium-grade, sandstone production centre; e.g. Wyoming, USA	144
	Total discoveries required to 2015	329*

*Some 50 of the total new mines are required due to depletion of deposits discovered in early part of the period 1977-2015.

Assuming an average 5-year lead-time between discovery and initial production, it was estimated that 329 new discoveries, representing a mixture of the six mine-types would be required by the year 2015, to provide the levels of production needed by 2020. Total "mine-life" reserves contained in these 329 new discoveries were estimated at some 9.5 million tonnes of uranium, and the total discovery cost (in 1976 dollars) was estimated at $50,000 million. In addition, the total capital investment (in 1976 dollars) required to provide the new production facilities was estimated at between $30,000 and $40,000 million. Figure I-7 attempts to provide a general impression of the rate of growth of required expenditures, both for these new sources of production as well as for the further delineation and development of known sources, over the period 1981 to 2020.

Thorium Resources Supply

The world's Reasonably Assured Resources of thorium, recoverable at costs up to $75/kg U ($30/lb ThO_2), are currently estimated at some 630,000 tonnes, almost 50% of the total being contained in monazite, in heavy-mineral beach-sand deposits in India. Much of the remainder is in similar beach-sand deposits, primarily in Australia, Brazil, Malaysia and the United States. These beach sand deposits are currently being exploited for their heavy-mineral content (i.e. minerals of titanium, tin and zirconium). Monazite is also recovered but primarily for its rare-earth rather than for its thorium content.

Present world production of thorium is in the order of 730 tonnes, all as a by-product of monazite. Should all of the thorium from the monazite production be recovered, it is estimated that up to 1,800 tonnes of thorium could be made available at current levels of production. Should additional thorium be required, it is estimated that world production levels could be expanded to at least 4,200 tonnes thorium a year, by recovering thorium primarily as a by-product of uranium producing operations in Canada, of copper mining in South Africa and of niobium mining in Brazil. Still further increases in production could be achieved by exploiting thorium as a principal product from thorite vein deposits in the United States. It seems unlikely, therefore, that thorium demand, even for nuclear purposes, will outstrip available supply from known sources by the year 2020.

Improvements in Fuel Utilization

Advances in technology are required if the utilization of our uranium resources is to be improved significantly. The estimates of uranium demand in this report are based on an enrichment tails-assay of 0.25%. New enrichment methods will likely be developed which may make it possible to reduce this tails-assay slightly, thus decreasing uranium demand. The maximum theoretical reduction in uranium demand would be 35%, should it be possible to reduce the tails-assay to zero. A further reduction in uranium requirements (up to 35%) could be more readily achieved by recycling uranium and plutonium in light water reactors. Similarly a reduction of about 50% could be obtained by recycling plutonium in heavy water reactors. The most significant advances in the technology of nuclear power, however, will arise from the ability to use more advanced fuel cycles.

The use of the thorium fuel cycle in thermal reactors and the plutonium-uranium cycle in fast breeders can reduce uranium requirements considerably but, as discussed below, the reduction obtained depends on the growth rate of the nuclear system. For this reason, the spallation reaction, as a means of producing fissile material from fertile material may become interesting in the future. All advanced fuel cycles require fuel reprocessing, active fuel fabrication and permanent waste storage in addition to the conventional "front-end" of the fuel cycle. Partly because of the long lead times involved, fuel enrichment and fuel reprocessing could serve as bottlenecks in the commercial development of nuclear power. Demands for these services in the short and long term are given in Tables I-8 and I-9, and illustrated in Figures I-8 and I-9.

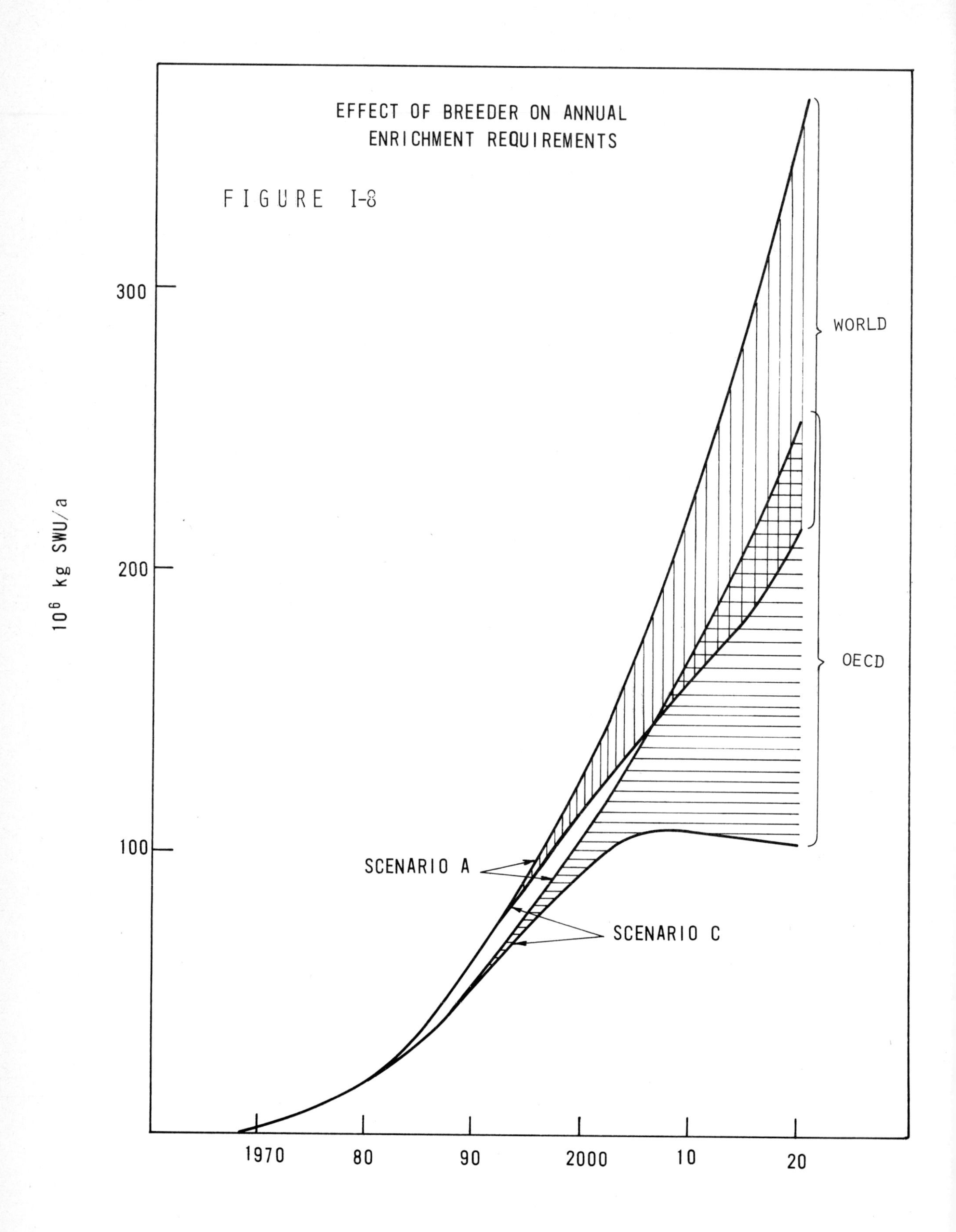
EFFECT OF BREEDER ON ANNUAL
ENRICHMENT REQUIREMENTS
FIGURE I-8
300
200
100
10⁶ kg SWU/a
WORLD
OECD
SCENARIO A
SCENARIO C
1970
80
90
2000
10
20

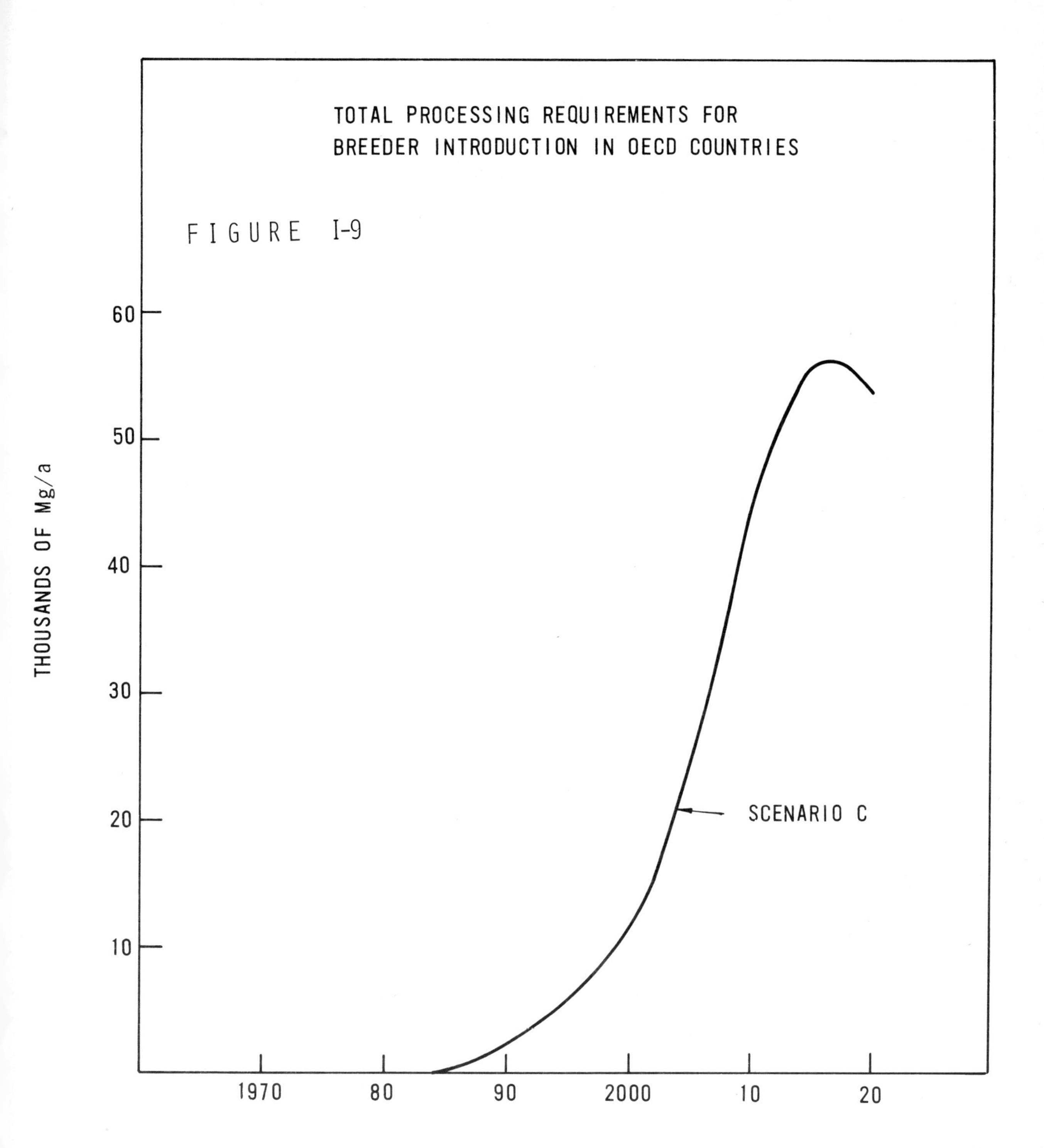
TOTAL PROCESSING REQUIREMENTS FOR
BREEDER INTRODUCTION IN OECD COUNTRIES
FIGURE I-9
THOUSANDS OF Mg/a
60
50
40
30
20
10
SCENARIO C
1970
80
90
2000
10
20

Reactors using these fuel cycles must also be developed. Except for the heavy water reactor, which can use the thorium cycle with little modification, most of the candidates for using advanced fuel cycles are in the prototype stage, the most prominent examples being the HTGR for the thorium cycle and the LMFBR breeder. The fuel cycle characteristics of these reactors must be improved if our uranium and thorium resources are to be more fully utilized.

Table I-8
ESTIMATED ANNUAL WORLD SEPARATIVE WORK REQUIREMENTS
(10^6 kg separative work units)

Nuclear Growth Scenarios	OECD (Regions 1-3)		Centrally Planned (Regions 4-5)		Remainder (Regions 6-11)	
	2000	2020	2000	2020	2000	2020
A	105	260	45	179	21	110
B	105	160	45	130	21	110
C	90	103	45	105	21	110
D	90	60	45	68	21	110
F	95	160	47	128	-	-

Table I-9
ESTIMATED ANNUAL WORLD FUEL-REPROCESSING REQUIREMENTS
(10^3 Megagrams)

Nuclear Growth Scenarios	OECD (Regions 1-3)		Centrally Planned (Regions 4-5)	
	2000	2020	2000	2020
A	-	-	-	-
B	2	68	0	56
C	11	53	4	31
D	10	35	3	30
F	13	65	5	40

Penetration of Advanced Fuel Cycles

Each nuclear growth scenario except scenario A calls for the introduction of new technology. The Fisher-Pry model was used to make an assessment of how rapidly this new technology can be introduced. Initially the growth of a new technology can be fairly rapid since it represents a small fraction of the total system. Thus, assuming that the breeder and the thorium cycle can be introduced as rapidly as TCR's were introduced about two decades ago, the initial construction rate for reactors using these technologies will be small, and will rapidly increase until it becomes the total nuclear plant construction rate. At this point, the penetration of the new technology slows down considerably, roughly to the nuclear system growth rate, since in a capital intensive system such as electricity generation, it is not normal to decommission plants prematurely in order to replace them with new plants. In this study it was found that breeder construction represented all new reactor construction after attaining 10-20% of the installed nuclear capacity.

Another possible limitation on the rate of penetration of advanced fuel cycles is the availability of fissile material. Nuclear reactors require fissile material for two purposes, inventory and operating requirements, the latter being proportional to the installed power and the former to the expansion rate. While inventory requirements for converter reactors are not negligible, they constitute a relatively small proportion of the reactor's total life-time requirements. The reverse is true for reactors using advanced fuel cycles with high conversion ratios. Indeed, if the conversion factor is equal to or greater than 1, inventory requirements are the only requirements for fissile material. Thus for systems consisting entirely of converter reactors, demand for natural uranium is primarily proportional to the energy generated, whereas for advanced fuel cycles, demand for fissile material arises primarily from inventory requirements, which depend on the expansion rate of the advanced system. The "asymptotic" reduction in annual uranium demand achieved by introducing several different advanced fuel cycles is shown in Figure I-10.

For high system expansion rates (10% or greater), which are characteristic of the initial introductory phase of nuclear power, the reduction in annual uranium demand obtained by introducing advanced fuel cycles is similar for all of the fuel cycles shown in Figure I-10. This similarity arises because the times required for the converter reactors to produce the inventory for the advanced cycle are about the same for each of these fuel cycle combinations.

Since the fuel-doubling time for breeder reactors is likely to be much longer than the system-doubling time initially, one expects that plutonium availability might have a constraining effect on the penetration rate of breeders, if they are introduced too early. It can be seen from Table I-10 that only in Western Europe are plutonium inventory values small and constant, thus implying that plutonium availability is limiting the rate of penetration of breeders into the Western European system.

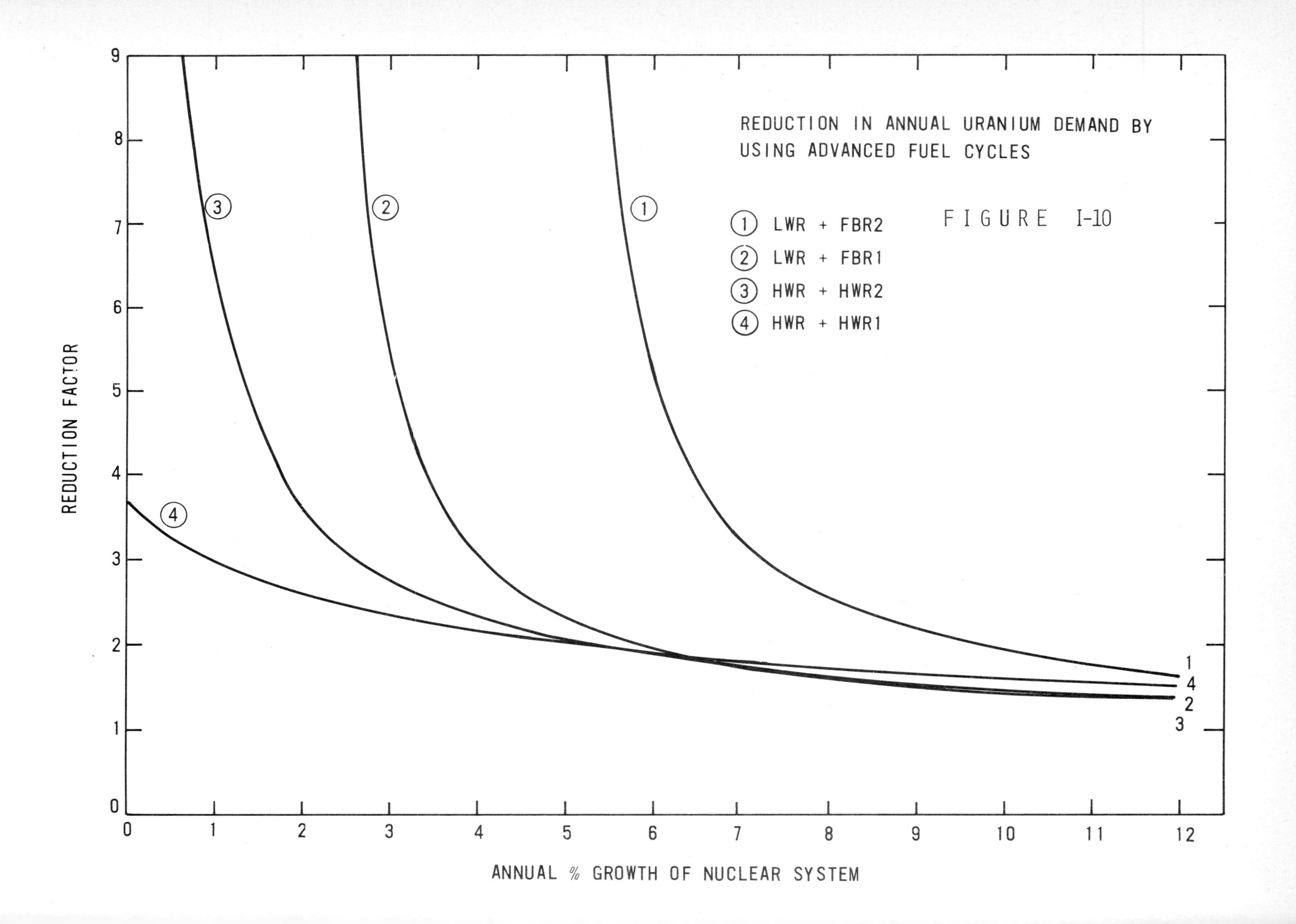
REDUCTION IN ANNUAL URANIUM DEMAND BY
USING ADVANCED FUEL CYCLES
FIGURE I-10
① LWR + FBR2
② LWR + FBR1
③ HWR + HWR2
④ HWR + HWR1
REDUCTION FACTOR
ANNUAL % GROWTH OF NUCLEAR SYSTEM
0
1
2
3
4
5
6
7
8
9
10
11
12

Table I-10
PROJECTED WORLD PLUTONIUM INVENTORIES
(10^3 Megagrams)

Nuclear Growth Scenario	North America		Western Europe		Japan		USSR	
	2000	2020	2000	2020	2000	2020	2000	2020
A	.8	3.4	.4	1.95	.25	1.14	.4	3.1
C	.7	.15	.03	.02	.22	.1	.25	.05

Principal Conclusions

The estimates of uranium and uranium enrichment requirements, deriving from this particular projection of nuclear energy supply, are expected to be on the low side for a number of reasons, not the least of which is the fact that the projections of electricity demand (Table I-1) are modest by historical standards. The assumption that, on the average, nuclear power will supply a maximum of only 50% of the electricity demand, leads to estimates of installed nuclear power which are lower than most other recent estimates. Moreover, only requirements for the production of electricity were considered. The use of nuclear energy as a heat source, for example, is expected to have applications in such areas as district heating, steel-making and synthetic fuel production, thus adding to the total demand for nuclear energy over the study period.

Perhaps one of the more important factors tending to depress the projection of uranium requirements is the assumption that the rate of penetration for advanced fuel cycles will be similar to that of converter reactors. This assumption is probably optimistic, since when light water reactors were initially introduced, certain key auxillary industries (e.g. enrichment) were already operating at a commercial level. The same situation does not exist with respect to breeder reactors and other advanced fuel cycles. Many areas of these new technologies will require more or less simultaneous development thus adding to the complexity of the overall program and to the likelihood that penetration rates will be lower than assumed.

With the single exception of the United States, the bulk of presently known uranium resources is distributed amongst countries whose nuclear programs are expected to be relatively small in the pre-2000 period. On the other hand, a rapid increase in nuclear power growth is foreseen after the year 2000 in regions which are expected to yield a large portion of the world's undiscovered uranium resources, thus possibly ameliorating the long-term supply situation.

Notwithstanding this potential advantage in the very long-term, the task of providing the world with uranium at the rate required, even under the rather optimistic demand assumptions of Scenario C, is truly immense. The problem is not likely to be one of existence of resources but one of availability of the required resources on an appropriate time-scale. Of the numerous factors which will make the task exceedingly difficult, the factor of time is perhaps the most crucial. Under the favourable assumptions of Scenario C, the industry is being asked to increase its level of production by fifteen-fold in less than 45 years, representing an average world rate of growth which is artificially low considering that the growth is required in different regions at different times. Indeed, no other segment of the mineral industry has been expected to accomplish such a task in so short a time. In view of the ever-increasing number of constraints and the lengthening lead-times required between initiation of new exploration programs and realization of initial production from successful discoveries, the 45-year period seems all too short.

The urgency of mitigating some of the various factors which constrain the expansion of world exploration and development efforts cannot be over emphasized. Without such encouragement the industry will have great difficulty in accomplishing a 15-fold increase in its level of production by the year 2020. Building up the required momentum to accomplish the task of meeting even the modest requirements projected under Scenario C seems problematical.

In view of the conservative aspects of the uranium requirement projections and the probable difficulties that will be encountered in meeting even these requirements, it seems clear that nuclear energy cannot succeed in supplying even 50% of the world's demand for electricity by the year 2020, without employing various of the advanced fuel cycles.

There are a number of important constraints, however, with respect to the introduction of the advanced fuel cycles, the most important of which is the need for a viable fuel-reprocessing industry. Even with the successful commercial establishment of these various new technologies, there will be limitations on the rate of their penetration imposed by the availability of fissile material. Indeed, in the post-2000 period, following the depletion of initial plutonium inventories, the effect of the introduction of the breeder will be marginal unless electricity growth rates fall below 4 or 5%.

In conclusion, although nuclear energy is expected to provide an increasing share of the world's total energy supply over the study period, its share of the electrical energy component will not likely exceed an estimated 50 to 60%. It is clear that providing even this fraction of future world energy supply presents a challenge calling for unprecedented levels of international co-operation. In this context, urgent action is required to promote the development of both uranium resources and nuclear technology.

UNCONVENTIONAL ENERGY RESOURCES

P. L. Auer
Sibley School of Mechanical and Aerospace Engineering, Cornell University

P. B. Bos, V. W. Roberts and W. C. Gough
Electric Power Research Institute, Palo Alto, California

UNCONVENTIONAL ENERGY RESOURCES

P. L. Auer[1], P. B. Bos, V. W. Roberts, and W. C. Gough
Electric Power Research Institute
Palo Alto, California (USA)

This report is concerned with the following unconventional energy sources: solar (which includes energy derived directly from sunlight as well as indirectly in the form of wind, waves, tides, ocean thermal gradients, or as fuel from biomass and other photochemical reaction products), geothermal, and fusion (otherwise known as controlled thermonuclear reaction). The theme common to these potential sources of commercial energy supplies is that they represent either renewable or virtually inexhaustible resources, and as the name unconventional implies, they are not now in widespread commercial use.

Interest in unconventional energy sources arises from the recognition that the commercial forms of energy in popular use today (liquid and gaseous hydrocarbons, together with solid forms of fossil fuels) represent limited and eventually exhaustible resources. As the world's population continues to increase and as people's aspirations for greater comfort and wealth lead to increasing per capita demand for energy, there is the fear that today's traditional supplies of energy will not be able to meet the demands of future decades.

It is not our role here to speculate on how long the world's supplies of petroleum, natural gas, coal, lignite, and so on, can or cannot continue as the principal forms of commercial energy, for that is done elsewhere in the Commission's study. Nor shall we discuss a form of solar energy in widespread use today, namely hydropower, which is also treated elsewhere. In a sense, our report may be considered a complement to the report on nuclear fission power, since between that resource and the ones we treat

---

(1) Sibley School of Mechanical and Aerospace Engineering, Cornell University, Ithaca, New York 14853 (U.S.A.)

are contained all the prospective long-term energy supplies of the world.

There are, however, a number of distinguishing features among the members of the category we have grouped under the title of unconventional. Some have reached fairly mature stages of technological development, but have yet to demonstrate economic advantages on a broad scale. The use of flat plate collectors to convert solar energy to low-grade heat for supplying hot water for residential use or heat for residential or commercial structures offers one example. Others, such as electricity production by vapor-dominated hydrothermal sources, suffer from an apparent resource limitation, although the technology is well advanced and is in commercial practice to a limited extent. Still others (the conversion of solar energy to electricity is an example) are known to be both scientifically and technically feasible but are not yet far enough advanced to judge whether they can become economically attractive in comparison with alternative means for generating electricity. Within the unconventional resource group, however, fusion still awaits the demonstration of its scientific and technical feasibility. Nevertheless, its future prospects may be considered to be rather promising, in an appropriately qualified sense.

In order to estimate the impact any or all of these resources may have on the world's energy economy within the next several decades (nominally to the year 2020) it would be necessary to estimate both the rate of technological advance in those cases where the current state of the art is less than mature and the rate of market penetration by those technologies that have reached a suitable stage of maturity during the interval under consideration. Needless to say, both of these are difficult judgmental tasks. Many studies have addressed these very same questions and have arrived at conclusions covering a broad range of quantitative estimates. We can claim no particular authoritative basis for our estimates or evaluations, other than that they were formed by groups of knowledgeable experts. We offer them here as a point of departure in the hope of arriving at as wide a consensus as may be possible for a category of technologies as relatively untried and unfamiliar as those contained within this group of unconventional energy sources.

In the main body of our report we discuss further each of the principal technologies and systems that are currently being considered as means for utilizing solar, geothermal, and fusion resources. Insofar as possible, their potential contributions to commercial energy supplies are discussed. It should be stressed, however, that in no sense do we attempt to present an additive compilation, for surely there will be competition among the emerging unconventional energy systems, as well as between them and the more traditional ones. At best, we merely attempt to place bounds on the very difficult and imprecise problem of future projections. In addition, we summarize briefly the potential resource base represented by each of the unconventional sources under consideration and the state of the art by which they may be harnessed in the future.

As a point of reference, one should bear in mind that estimates of the current global use of commercial energy are approximately 250 EJ per annum (250×10^{18} J/yr). One may convert this figure to units of power by considering the annual average rate at which energy is used, whereupon the global figure becomes approximately 8 TW (8×10^{12} W). It is also estimated that among the rural population of the developing nations the rate at which noncommercial forms of energy (principally wood and animal dung) are used may be as high as 1 TW. Based on a world population of 4×10^9, the per capita use of energy is at the rate of 2 kW, but the geographic distribution is highly uneven. In North America the per capita figure currently exceeds 10 kW, while in western Europe it is 5-6 kW, and for many developing nations, it is 1 kW or less. Even larger imbalances exist between the rural and urban populations of many developing nations.

The role of unconventional resources, therefore, may be expected to vary not only as a function of which particular category is under consideration but also as a function of which geographical market one wishes to examine. Since unconventional resources will initially enter in a substitutional mode, it is after all the rates of market capture that will determine the magnitude of their contribution. This, in turn, is governed by relative economic factors, and one may expect some variation in these factors as one moves from one geographic region to another.

By far the largest source for potential growth in energy demand lies with the developing nations, which now account for more than half the world's population, have population growth rates that are three times higher than those of the industrialized nations, and whose per capita energy use lies well below the world's average value. As these nations progress along the road of development, certain unconventional resources may play a more significant role (and at an earlier date) than may be the case with developed nations.

We shall now turn our attention to three categories of energy sources: solar, geothermal, and fusion. The reader is cautioned, once again, that we are attempting no more than to place some realistic limits on how and to what extent each of these sources may contribute.

SOLAR ENERGY

THE SOLAR RESOURCE

The radiant energy arriving on earth from the sun represents by far the most prodigious source of energy available to humanity. The solar flux at the outer fringes of the earth's atmosphere is 1.35 kW/m^2. A disk having the size of the earth's diameter, then, intercepts 178 000 TW, or 20 000 times more power than the 8-9 TW that humanity expends on its organized activities. Of course, it is this vast amount of energy that makes our planet habitable, accounts for our climate and the bulk of our food requirements, and much more.

Only a portion of this solar energy, however, arrives on the surface of the earth. The atmosphere acts as both an absorbing and a reflecting medium with somewhat unpredictable, turbulent, scattering properties. In favorable locations, the maximum peak solar power density reaches 1 kW/m^2 at sea level, and the maximum average power density is about 250 W/m^2 (global insolation) and 400 W/m^2 (direct-beam radiation at normal incidence). Significant geographic variations are also to be expected, as indicated by Table I.

Table I

AVERAGE TOTAL IRRADIANCE (insolation)

Location	kWh/m^2-day	W/m^2 (average)
Tropics, deserts	5-6	210-250
Temperate zones	3-5	130-210
Northern Europe	2-3	80-130

In addition to diurnal variations, there are the seasonal variations, which may range from negligible in the tropics to as much as a factor of 10 in the higher latitudes, with northern Europe as an example. Climatic variations must also be considered. For these reasons it would seem that widespread use of solar energy will entail either the extensive use of secondary energy storage (possibly including an energy transport and distribution network) or hybrid use in conjunction with other sources to permit a reliable, noninterruptible energy supply. Nevertheless, the attractiveness of sunlight as an energy source, due to its high thermodynamic quality (the effective blackbody temperature of the solar corona is 5900 K), global abundance, and renewable and nondepletable character, would indicate that it will eventually become an important contributor to commercial energy supplies.

Direct manifestations of solar energy in the form of direct-beam or diffuse radiation may be utilized to produce heat or secondary forms of energy, such as electricity or synthetic fuels. Indirect manifestations of solar energy, as they appear in the wind, waves, ocean-thermal gradients, photosynthesis, and falling water may also be harnessed usefully. While we do not treat falling water (hydropower), we do include in this category the tides, which properly speaking are a manifestation of lunar energy rather than solar.

SOLAR-GENERATED THERMAL ENERGY

Solar energy in the form of direct-beam radiation lends itself effectively to the production of heat. Temperatures attainable depend on the design of the collection device, limited only by the effective radiation temperature (see above), with heat at less than 100°C readily achievable without concentration (concentration ratio of unity). The maximum temperature attainable is approximately 5600°C, but no useful work would be obtained in this extreme. The solar furnace at Odeillo, France, has achieved temperatures approaching 3000°C with concentration factors in excess of 10 000.

Low-Temperature Heat

The generation of low-temperature heat (less than 100°C) from sunlight is a fairly mature technology with a long history of application in many parts of the world. Modern developments have led to certain improvements in efficiency of conversion and system design. The basic system today, as previously, consists of a flat plate collector, liquid or air heat transfer medium, with or without storage. It finds applications in water heating and space heating. In general, air conditioning requires somewhat higher operating temperatures.

Water-heating systems can have annual net efficiencies of 30-40% and will cost between $\$100/m^2$ and $\$200/m^2$, uninstalled. (Current US$ are given unless otherwise noted.) Application to space heating and cooling can be assessed properly only if both passive measures and active solar systems are considered together. A wide range of passive, energy-conserving architectural and engineering techniques are available, which can lead to significant reductions in energy requirements. Costs for implementing passive measures through suitable design, the use of appropriate materials of construction, insulation, and so on, are far less expensive than active solar devices and can reduce annual thermal needs by as much as 50% when compared with many existing structures.

Only after such measures have been taken does it make economic sense to consider active systems. Nearly all such applications use flat plate collectors, water or rock energy storage, and a heating, ventilating, air conditioning system. Active solar systems can provide as much as 50-70% of total thermal requirements, with an annual utilization factor of 35-20% (decreasing as the solar share of total increases). Minimum costs of currently available collectors are in the range of $\$150/m^2$ to $\$200/m^2$; commercially installed systems have cost as much as $\$500/m^2$.

Total economic costs, of course, will be determined by a variety of factors (for example, taxes, financing charges). One contribution to higher cost is the requirement for some form of backup--electric, gas, or another form of heating. Interfacing with the grid will present problems insofar as the solar-equipped user can degrade the local utility's load factor. Load smoothing may be achieved through increased storage capacity or by incorporating integrated storage and a heat pump and allowing the solar system to provide thermal energy to the reservoir from which the heat pump draws energy. The overall economic benefits of solar-assisted heat pump systems have yet to be fully determined.

We estimate that simple applications of solar energy to water heating and space conditioning are unlikely to displace more than 2-5% of the world's primary energy requirements over the course of time, and it is unlikely they will reach this range by the year 2020. Much of the world's population lives in regions where large quantities of heat for such applications are not required, while in those regions where current requirements of primary energy for this purpose are large, vigorous conservation measures should be highly effective in reducing the demand. Consequently, the potential market where solar heating and cooling can provide 50-70% of the thermal load is reduced considerably.

Solar-Thermal Electric Conversion

A promising technology incorporating solar-generated thermal energy is the solar-thermal electric conversion (STEC) system for the production of electricity. The same technology could also be used effectively to produce the higher-grade heat needed for a variety of industrial processes. The basic elements of STEC involve an optical concentrator to gather solar radiation, an absorber for the thermal energy collected, a heat transfer system, and a more or less conventional balance of plant for generating electricity. In principle, STEC power plants can come in a variety of sizes, ranging from a few hundred megawatts (electric) down to far smaller-size units.

The basic principles are well known, with several solar-driven hot-air and steam engines having been constructed between 1880 and 1920, the largest of which was a 45-kW installation in Meadi, Egypt. The 1-MW (thermal) furnace at Odeillo, France, has been employed more recently to give further proof of concept information and operating experience.

An active development program on STEC is being pursued in the United States, with plans for the construction of units varying in size from 5 MW (thermal) (near completion in New Mexico) to 10 MW (electric) (near Barstow, California) to a 100-MW (electric) unit to be constructed in the mid-1980s. Two private firms, Ansaldo (Italy) and MBB (Federal Republic of Germany), have joined to offer STEC units based on Professor Francia's design in sizes up to 1 MW (electric). Additional construction plans for STEC systems have recently been announced by France.

A typical 100-MW (electric) STEC unit might be envisioned as consisting of 12 500 heliostats, each having a reflecting surface of approximately 40 m^2, with a central receiver tower 250 m in height, supporting an absorber to provide steam or hot gas to a turbine for periods ranging from six to eight hours per day. Present designs incorporate either conventional fuel storage for operation of the plant in a hybrid mode or thermal storage sufficient to extend operation to the intermediate power-generating mode. Overall efficencies are expected to be in the range of 15-20%. Cost estimates for STEC systems with central receivers range from $1000 to $3500/kW (electric), while estimates for distributed collector concepts range from $2000 to $5000/kW (electric). Such estimates, however, are based not on actual construction experience but on the expectation of certain achievements from the demonstration program. The largest unknown, perhaps, is the cost of heliostats, and this may not be accurately determined until a program for mass production is in place.

Reliability of energy supply by means of STEC systems will be a function, in part, of the siting strategy chosen and the nature of the total utility system with which STEC units would be interconnected. By the end of the next decade, there should be sufficient experience with operating units in the range of 50-100 MW (electric) so that a meaningful estimate of STEC's economic attractiveness can be made. The use of solar thermal electric conversion plants for base load generation, however, has received relatively little attention so far.

PHOTOVOLTAIC ENERGY CONVERSION

Solar radiant energy may be converted to direct-current electricity, typically by means of suitably constructed thin layers of silicon or other semiconducting material. The silicon solar cell, developed in 1955 by the Bell Laboratories (USA), has been the mainstay of spacecraft power systems for the past two decades.

The potential advantages of photovoltaic conversion are impressive. There need be no moving parts; lifetimes can (in principle) exceed 100 years, although cell performance may be expected to degrade continuously over its operating lifetime; maintenance involves little skill; both direct-beam and diffuse solar radiation can be utilized effectively; the system is inherently modular and readily lends itself to the design of virtually any system size, small or large.

Present costs of solar cells are approximately $15,000 per peak kilowatt of electricity, with proven efficiencies of 12-15%, as compared with efficiency limits of about 23%. Consequently, the cost of producing electricity by this means is some 100-200 times more expensive than by conventional methods.

Current research and development efforts are concentrated on devising manufacturing techniques that may reduce the fabrication costs by large factors, with $500 per peak kilowatt the target figure for 1985. Alternatives to the silicon cell, CdS-CuS being one example, GaAs (which lends itself to higher efficencies when used with concentrators) being another, may provide for lower overall solar cell costs. The wide-scale use of photovoltaic devices for power generation will require not only major cost reductions in the fabrication of the cells but also proven reliability and compatibility with the remainder of the electric grid system; economic and reliable energy storage; and high-power, solid-state regulators and inverters for the conversion of dc current to stable ac current. Both the promises and the challenges are formidable. If either or both STEC and photovoltaic technology can begin to enter the electricity generation market economically by the 1990s, a significant fraction, perhaps 10-15%, of the electricity produced globally by the year 2020 could be furnished by these means. One must note, however, that at this time such estimates are highly conjectural.

SOLAR HYDROGEN PRODUCTION

One means to utilize solar energy, in principle, to furnish a major portion of the world's energy needs is to find a suitable secondary form of energy that can be readily produced via solar energy. Electricity is a likely candidate, but suffers at the moment from our inability to store large amounts of it economically or to transport it over arbitrarily large distances. Hydrogen gas is another potential candidate, and some of its proponents consider it or some derivative of it (for example, methanol) to be the most likely candidate for the future.

A solar-hydrogen-dominated energy economy envisions large solar installations in the sunny, arid regions of the world, which in themselves are inhospitable sites for any other human activity. These would be coupled by pipeline and tanker transport and distribution networks to the world's energy markets. Large-scale storage could employ various naturally occurring geologic formations or, alternatively, storage in the more expensive liquified form. Of course, hydrogen could be produced from primary energy sources other than solar and in this way could decouple the user's needs for secondary energy from the nature of the primary source.

Hydrogen can be produced by thermochemical means, by photolysis, or by electrolysis. Experience to date on the first two of these methods has not been very encouraging, but so far relatively little serious effort has been devoted to the matter. Present estimates, admittedly crude, for the cost of producing hydrogen by solar-thermochemical means range from $40 to $80 per barrel of oil equivalent. The lower end of the spectrum approaches the cost of heating oil in Europe; however, that figure includes taxes as well as the costs of transport and distribution. It would be premature to estimate at this time the potential contribution of hydrogen to the world's energy needs for the year 2020, but the matter certainly deserves continued attention, together with an appropriate research and development effort.

INDIRECT SOLAR ENERGY

Wind Energy

Until displaced by the steam engine, wind energy was used on a large scale for commercial and agricultural purposes in England, Germany, France, Denmark, and Holland. Current development efforts are oriented toward electricity generation, using large individual wind generator units, for distribution through existing electric networks.

Wind power scales as the cube of the wind velocity (velocity raised to the third power) so that appropriate siting in favorable locations is a primary consideration to its use. Conversion efficiencies can be quite good, in the range of 20-40%. A wind generator located at a site with mean windspeeds of 7 m/s would intercept a power density of approximately 200 W/m^2, of which 40-80 W/m^2 could be converted to electricity. (By comparison, a favorable site in terms of solar insolation may be expected to produce 40-80 W of electricity for each square meter of collection area.) In general, however, the amount of land area required by the wind generator to produce a given amount of electricity will be four to five times greater than that required by a solar-electric converter, although it may be possible to use the land for multiple purposes (agriculture, for example).

Various authors have estimated the total energy contained in the global atmospheric circulation (on the order of a few hundred terawatts), but the number is not relevant for our purposes. What one wishes to know is the windspeed and its duration at 20-30 m above ground level at various locations throughout the world. There is a paucity of such data. By comparison to direct solar energy, the amount of energy potentially available from the wind is probably only a few percent of the former.

Most wind generator development work in the United States is concentrated on two-blade, propeller-type units mounted on a horizontal axis. A test facility consisting of a 38-m (125-ft) diameter swept-circle rotor is operating in Ohio. Two additional units of comparable size, driving 200-kW (electric) generators, are scheduled for installation in 1977. Two larger machines, also with two-blade propellers, 61-m (200-ft) diameter swept-circle, driving 1500-kW (electric) generators, are to be installed during 1978 and 1979. An experimental 91-m (300-ft) diameter unit is under design.

In Canada, a 37-m (120-ft) high Darrieus (vertical axis) rotor is being constructed and is nearly operational. It is the largest of such devices, coupled to a 200-kW (electric) generator, and is one of the few major projects departing from the horizontal-axis approach that most other countries are following.

Economic considerations are likely to limit use of wind generators to locations with mean windspeeds greater than 5 m/s. In addition, such factors as diurnal and seasonal wind variations, accessibility to the site, problems associated with interconnection to the remainder of the utility network, adaptability of loads to interruptible supply, and the question of load scheduling are but some of the issues that must be properly evaluated before one can make a meaningful assessment of wind power's future role. At the present, the simplest mode for the use of wind power on a substantial scale is by connection to a grid with existing hydro capacity, where energy storage may be accomplished by displacement. Wind generators today cost two to four times more than conventional generators of equivalent rating. Nevertheless, wind power may become a significant energy source on a regional basis where conditions (windspeeds, hydro capacity) are particularly favorable.

Wave Energy

Historical attempts to exploit the power of waves for human purposes have been relatively few, in spite of the extensive human experience with ocean waves over the course of thousands of years. Wave energy derives from wind energy, which in turn derives from solar energy. By comparison with the solar energy, however, the amount of energy stored in waves is not nearly as enormous, having been estimated to represent about 2.7 TW globally.

A more accurate and somewhat more relevant measurement at a site in the northeast Atlantic off Scotland (59°N 19°W) indicates that the annual power crossing a line perpendicular to the average propagation direction of waves is 91 kW/m. Taking this number as representative for the larger, neighboring area, one would estimate the wave power potential for western Europe, for example, as 2×10^7 kW (electric) on the average. Here we assume an overall conversion efficiency of 20%, although it might reach as high as 40% or as low as 10%. The 2×10^7 kW (electric) generating capacity, leading to the production of 1.75×10^{11} kWh of electricity annually, could be realized, in theory, from a series of interconnected strings of converters, some 100 in number, each about 10 km in length. This represents, of course, an immense undertaking involving the northern European coastline.

The engineering challenge is to find a cost-effective way to convert a fraction of the large amount of energy contained in wave motion into a mechanical, hydraulic, or pneumatic form that may be used to generate substantial amounts of electric power, under the inhospitable conditions existing on the open seas. Since waves come in a wide range of wavelengths and amplitudes, any effective device either will have to be broadband (nonresonant) or will have to have its resonance frequency continuously adjusted. Many mechanisms have been proposed and a number are being investigated. For example, waves can be made to compress air in the top of a floating tank to a pressure sufficient to drive a pneumatic turbine. Waves passing a standing pipe equipped with a flap valve can raise the water level in the pipe to many times the wave height, making it possible to drive a conventional hydroturbine. Another class of mechanism employs two floats with different dynamic response characteristics. The differential motion is used to operate a pump or a mechanical engine. For example, a long vertical cylinder extending far below the surface will remain nearly stationary as the waves pass by. A toroidal float surrounding it will rise and fall with the waves and can be made to push and pull on pump plungers attached to the cylinder. Another differential motion concept, called the nodding duck, employs large cam-shaped "ducks" mounted on a very long floating frame. The ducks oscillate, or nod, relative to the frame, driving hydraulic pumps.

Worldwide application of wave energy conversion systems will be limited by geographic factors and the economics of energy transport and distribution over large distances. It is difficult to estimate the potential contribution of this source, although it might be comparable to that of land-based wind-powered systems.

Tidal Energy

Two tidal stations operate commercially at this time, one at La Rance in France and a small experimental unit in the Kislaya Inlet of the USSR. Although discussions of the prospects for the utilization of tidal power have extended over many years, the total resource base is small. It has been estimated that no more than 25 sites, worldwide, meet the necessary conditions of tidal amplitude and coastal topography. Ignoring engineering limitations, this represents an average hydraulic power of about 0.06 TW, of which no more than 10-25% could be converted to electricity. In practice, it is unlikely that more than a few percent of the total would ever be harvested usefully. Table II lists the sites that have been identified to date.

Table II

TIDAL POWER SITES

Location	Average Tidal Range (m)	Hydraulic Energy (10^9 kWh/yr)
North America		
Bay of Fundy		
Passamaquoddy	5.5	15.8
Cobscook	5.5	6.3
Annapolis	6.4	6.7
Minas-Cobequid	10.7	175.0
Amherst Point	10.7	2.25
Shepody	9.8	22.1
Cumberland	10.1	14.7
Petitcodiac	10.7	7.0
Memramcook	10.7	5.2
Cook Inlet, Alaska		
Knik Arm	7.5	6.0
Turnagin Arm	7.5	12.5
South America		
Argentina		
San Jose	5.9	51.5
Europe		
England		
Severn	9.8	14.7
France		
Aber-Benoit	5.2	0.16
Aber-Wrach	5.0	0.05
Arguenon/Lancieux	8.4	3.9
Frenaye	7.4	1.3
La Rance	8.4	3.1
Rotheneuf	8.0	0.14
Mont St. Michel	8.4	85.1
Somme	6.5	4.1

Table II (continued)

USSR		
Kislaya Inlet	2.4	0.02
Lumbovskii Bay	4.2	2.4
White Sea	5.7	126.0
Mezen Estuary	6.6	12.0

Source: M. K. Hubbert. Chapter 8 in Resources and Man, W. H. Freeman (1969), except for Knik Arm and Turnagin Arm, which were taken from Tidal Power, ed. T. J. Gray and O. K. Gashus, Plenum Press, New York (1972).

Notes:

1. The basin area depends on the position of the dams and reservoirs and for some locations the best position is not obvious. Thus the energy figures are open to some variations.

2. The sites have been screened for tidal range and general engineering feasibility but not for economic feasibility in terms of distance from energy load centers, comparison with other local energy sources, and so on.

3. It is likely that there exist tidal power sites other than those listed but not large ones near heavily electrified regions.

4. Of those listed, only La Rance and Kislaya have been developed.

Ocean-Thermal Energy Conversion

The concept that the temperature difference between the surface and the deep waters of the ocean could be used as a source of energy reaches back nearly a century. Recently, interest in this option was revived, and detailed technical and economic analyses have been carried out. The general conclusion from these reports is that major technical problems and unfavorable economics make ocean-thermal energy conversion (OTEC) systems relatively unattractive in the total perspective of solar energy options.

It has been estimated that by placing OTEC facilities approximately 15 km apart throughout the oceans between 20°S and 20°N latitude, the theoretical maximum limit for electricity production would be 50 TW (electric), equivalent to 150 TW (thermal). An ultimate installed capacity representing a small fraction of this potential would still represent a significant contribution to the world energy production needs in the next century, hence the considerable interest in the OTEC option. Nevertheless, we believe the engineering problems are so complex, with little margin for maneuver, that OTEC's potential must be considered as somewhat speculative.

The OTEC plants that have been investigated used a closed Rankine cycle incorporating a working fluid suitable for the temperatures and the temperature differences that would be encountered. Warm seawater would be pumped through an evaporator, and the working fluid would be vaporized. The vapor would expand through a low-pressure turbine, creating shaft horsepower that could then be used to drive an electric generator. The vapor would then be condensed by thermal exchange with deep ocean water and brought to the condenser by the cold-water inlet pipe. The working fluid would then be pressurized and pumped to the evaporator to continue the cycle.

A large number of technical and economic problems are associated with the development of OTEC systems. Some could be resolved through further research and development. These include improving the efficiency and lowering the costs of the heat exchangers, finding a solution to potential environmental hazards created by the large amounts of ammonia used in the system, and examining alternative designs for the outer hull of the plant structure. However, other problems are inherent in the system itself. Essentially a tremendously large system--one consisting of pumps, heat exchangers, and other expensive equipment--would be required to generate a relatively small amount of power. Because of the hostile marine environment, significant problems of materials, operation, and lifetimes would be encountered. In addition, the power generated would have to be economically transported to land-based load centers. In general, ocean-thermal energy conversion technology does not promise technical and economic viability comparable to the solar-thermal conversion and photovoltaic systems discussed earlier.

Market Penetration of Solar Energy Systems

The development and large-scale commercial diffusion of new energy technologies take considerable time. Each technology, whether it is the light water reactor or a solar-thermal electric power plant design, must move through a sequence of stages from demonstration of scientific feasibility to widespread and significant use. It may be assumed that the four decades characteristic of the sequence of events for the light water reactor can also be expected for other technologies.

While it is obviously not possible to predict with certainty the rate and scale of diffusion of any technology, it is possible to estimate the maximum rate of penetration under the conditions that the technology is commercially developed and economically, environmentally, and socially attractive.

An optimistic scenario for market penetration in the United States by a mix of solar-energy options assumes that 1% displacement could begin by 1985 and grow to 50% by 2050. In order for solar energy to achieve this 50% takeover in a 75-year period, the rate of fractional penetration would have to be higher than that which occurred historically for natural gas. Yet solar energy systems are considerably more expensive and complex than the natural gas infrastructure was. It would be remarkable if solar options could really penetrate the total energy marketplace at such a rate.

The important point is that even if this could happen, it would require the better part of a century for a mix of solar-energy systems (assuming technical success, economic competitiveness with the other energy options, and total social acceptability) to displace 50% of the United States energy demand. This again highlights the need for an aggressive and sustained program to develop a wide range of advanced energy systems options so that the best of these can be widely used by the time the fossil sources have been largely exhausted.

The commercial applications of solar power are at such an early stage of development at this time that any forecast of their potential contribution within the next 40 to 50 years can be based on little more than guesswork. Having stated these reservations, we nonetheless have made a series of "educated guesses" in order to arrive at the forecasts shown in Table IIa. In order to appreciate the relative magnitude of potential contributions from solar energy in the time frame of 1985 to 2020, the reader is reminded that the current annual average rate of energy use, worldwide, is in the range of 8-10 TW and that it is likely to increase to a range of 30-40 TW by 2020.

Table IIa

RENEWABLE ENERGY SOURCES

Source & Application	Global Abundance TW[1]	Theoretically Recoverable TW[2]	Range of Contribution by Year 2020 TW[3]
SOLAR	178,000	50 - 100	
Heating & Cooling			0.5 - 2.0
Electricity			0.1 - 0.3
Fuel			?
WIND	350	?	
Electricity			< 0.01
Mechanical Work			< 0.2
WAVES	3	0.1 - 1.0	
Electricity Fuels			< 0.01
TIDES	3	0.06	
Electricity			< 0.01
FUSION			< 0.05

Notes: (1) Crude Estimate
(2) Assuming Appropriate Technology Can Be Developed Economically
(3) Assuming Current Projections for Near-Future Technological Successes are True

SOLAR ENERGY IN DEVELOPING NATIONS

Bioconversion Systems

The primary photosynthetic production rate of the world's forests is perhaps 50 TW (thermal), roughly ten times greater than the total present production of oil and natural gas. Until recently, the conversion of sunlight to energy in the form of wood was the major energy source in much of what is now the industrialized world. Up to 1 TW (thermal) in the form of firewood, waste, and animal dung may presently be in use among the rural population of the developing nations. Therefore, one may well ask whether biomass can represent a potentially major energy resource in realistic terms.

There is an urgent reason for considering the use of biomass production as a fuel source for the developing countries. There is presently a worldwide crisis in firewood--it is, in fact, the "other" energy crisis, and it is far more real and difficult than the "oil crisis." Increasing demand for firewood has caused increased pressure on forests, and deforestation on a globally significant scale may well occur. In much of the developing world, firewood is now regarded as one of the major costs of living. Such pressures make replanting or forestation campaigns impossible; trees are often torn out soon after they are planted.

Moreover, combustion of dung and agricultural waste robs the soil of important nutrients and of badly needed soil structure. The process leads eventually to a dead end as both firewood and soil fertility are exhausted, with no means to replenish them.

A thoughtfully designed plan for biomass conversion offers the potential of turning this destructive pattern into a regenerative cycle. The production of alcohol, for example, can lead to the production of both fertilizers to revitalize the soil and fuels to replace the use of firewood.

The sources of biomass for energy conversion are numerous. Sugarcane, an unusually high-yield plant, is potentially very important in this regard. The growth of a class of plants, euphorbia, which are closely related to the rubber tree and produce a hydrocarbon somewhat similar to petroleum, has been proposed.

An optimistic estimate of the costs of establishing energy plantations and operating such a system has been made: fuels could be produced for a cost of approximately $1/GJ, considerably less than the present world market price for oil.* Such fuels could be used as combustion sources for power plants, or they could be transformed by a number of biochemical processes to liquid fuels (such as methanol) or gaseous fuels (such as methane). Subsequent conversion to liquid fuels would be necessary for markets other than power generation.

In another study,** an analogous analysis for the production of biomass from trees and sorghum to be used in a thermal power plant has been carried out. However, the conclusion was that such an operation would be far more expensive than the estimate given above. Estimates here are $2530 and $6000 per kilowatt (electric) of rated capacity (load factor = 0.8) for use of sorghum and sycamore trees, respectively. In addition, annual operating costs could vary substantially. The economic advantages of fuel plantations are still very much of a debatable issue.

In general, photosynthetic conversion efficiencies are very low. The most productive sugarcane systems have net conversion efficiencies in excess of only 3%. That is the highest of all the systems discussed here. In contrast, euphorbia would have a conversion efficiency of about 0.5%.

In the industrialized countries, biomass conversion could, in principle, be a substantial source of energy where adequate land is available. Others have estimated that the worldwide production of fuels from biomass could be as much as 10 TW (thermal). If only 50 km^2 are required for each 100 MW (thermal), this would correspond to an area of 5×10^6 km^2 dedicated to continuous biomass production.

However, in both the developing and the industrialized nations, the already strong pressures on land for the production of food, fiber, and forest products will only intensify in the future, and practical concerns may limit the ultimate production of fuels from renewable biological sources well below their potentials.

---

*George Szego. "Design, Operation, and Economics of the Energy Plantation." In Proceedings, Conference on Capturing the Sun Through Bioconversion. Washington, D.C.: Center for Metropolitan Studies, March 1976, pp. 217-240.

**See, for example, R. E. Inman, et al, "Silvicultural Biomass Farms" Technical Report No. 7347 (6 vols.), The MITRE Corporation (May 1977).

Even production of useful and efficiently combusted biomass fuels at the rate of 0.2 TW (thermal) would be important, however, since this could displace the 1 TW (thermal) corresponding to the use of firewood, dung, and wastes in the developing world. Others, however, argue that levels approaching 5 TW (thermal) are possible. While the estimate appears highly optimistic and the scale of production seems formidable, movement in this direction may be an important alternative to the present trend toward worldwide deforestation.

Systems Considerations

The question of to what extent renewable resources may play a significant role in the future energy markets of the developing nations can be singled out for special consideration. Among the developing countries much of the energy used, such as firewood, dung, and agricultural wastes, is converted to useful work at far lower efficiency than are the primary resources of the industrialized world. Normally, these fuels are converted to useful energy--for cooking, heating, irrigation, and other agricultural tasks--with efficiencies of 3-5% compared with around 20-30% for modern fuels and energy-using equipment. Paradoxically, there is hope here. Once the shift can be made to using energy more effectively through better sources and user equipment, large increases in well-being and productivity can (in theory) be gained from quite small extra inputs.

The traditional discussion of the use of solar energy in developing parts of the world has centered on the identification (or attempted identification) of local markets for the use of small-scale technologies to harness wind, running and falling water, and solar radiation to provide energy for irrigation, cooking, lighting, hot water, and, possibly, manufacturing and communications.

In agriculture, perhaps the most critical resource-based factor in the developing world, energy is now the central issue. The need for energy for pumped irrigation and the production of fertilizers alone requires a substantial increase in efficient use of a far greater amount of energy than is presently available to the developing world.

At the same time, large amounts of energy are being expended on the generation of lighting and heating. For example, in India nearly half the total energy now consumed is for lighting and cooking. In the year 2000, it will still be a major problem to provide feasible options to meet the domestic needs of the rural population, which may be 650-800 $\times 10^6$ by the year 2000, spread over 500 000 villages. Even then, one-fifth to one-third of the nation's energy requirements will be for cooking and lighting.

According to some research reports, biogas production from animal dung may be the most important technology for converting biological material to other, more useful forms of fuel. It has the advantage of producing a residue that is an excellent fertilizer, and the gas, of intermediate quality (20 000 kJ/m^3), can be used for lighting and cooking more efficiently and in a more healthy manner than direct combustion of animal and agricultural waste. By combining cooking and electricity production, the estimated cost of electricity from biogas is in the $0.05-$0.08 kWh (electric) range. (This corresponds roughly to the cost of electricity produced from photovoltaic systems with an installed cost of $1000 to $2000/kW [electric] in regions of high insolation.)

Some 200 $\times 10^6$ well-fed bovine animals could, in principle, provide three-quarters of the domestic needs of an estimated 800 $\times 10^6$ Indians living in rural areas in the year 2000 through 1.5 $\times 10^6$ community biogas systems. In such a scheme, the rural requirements for kerosene could be completely eliminated. In addition, the plants would produce 5 $\times 10^6$ tons of nitrogen per year in the form of residue.*

A recent and detailed technical and economic analysis has been carried out for the use of conventional and renewable (solar, wind, biomass) energy sources for prime movers for small (3.7 kW and less) irrigation pumps.** Because mechanized irrigation is one of the most important needs of the developing world, and because in some regions of the developing world, such as India, over two-thirds of the commercial energy for agriculture is used to power

---

*K. Parikh. "India's Fuel Needs and Option." International Conference on Energy Resources. International Institute for Applied Systems Analysis, Laxenburg, Austria, 1976.

small diesel and electric irrigation pumps, a careful analysis of the options available for irrigation from renewable sources is warranted.

The study indicates that the community-size biogas plants would be the least expensive renewable source already competitive with current fuel oil prices ($0.10 to $0.25/liter) in most countries. However, the development of small, commercial engines operating on intermediate-grade biogas, with costs similar to present diesel engines, is required to permit effective use of biogas for irrigation.

The study also concludes that for fuel prices in the $0.40-$0.50/liter range, solar electric photovoltaic pump sets will become competitive when solar array prices drop by approximately a factor of 100, that is, to between $250 and $300 per kilowatt. Many of the attributes of photovoltaic energy systems may be of particular interest to developing countries. Specifically, their modular quality, simple maintenance, and ability to operate with diffuse sunlight would allow them to provide much of the energy required in a form uniquely suited to high efficiency in end use, without making difficult demands on available land areas.

Of course, small-size solar-thermal-powered irrigation pumps are already in use in certain areas of the world where sunshine is plentiful. France has produced commercial units, in sizes ranging up to 25 hp (19×10^3W), which use flat plate collectors and a low-boiling organic fluid to drive a Rankine-cycle heat engine. An extensive development program exists in the United States. A 25-hp (19×10^3W) unit, using a parabolic trough collector and an organic fluid Rankine-cycle heat engine, is operating in New Mexico (USA). A 200-hp (150×10^3W) version of similar design, but coupled to an electric-drive pump, will be built in Arizona (USA) in 1979. Smaller units, using either photovoltaic devices or wind turbines, are also being tested in the United States. Based on current experience, even the simplest solar-thermal irrigation units cost in excess of $10 000 per horsepower and must come down in cost by an order of magnitude before a large market interest can be expected.

**V. Mubayi and T. Lee, "Irrigation in Less Developed Countries — A Study of the Comparative Costs of Using Conventional and Renewable Energy Sources for Powering Small Irrigation Pumps in Developing Countries", Draft Report, Policy Analysis Division, National Center for Analysis of Energy Systems, Brookhaven National Laboratory, Upton, New York 11973 (USA) March 1977.

For many crops the peak water demand at certain times of the year is significantly higher, sometimes by a factor of 2.5 (rice and sugarcane), than average demand. In these cases, the most economical systems may combine solar cell units with supplementary diesel fuel. In such hybrid applications, the diesel sets would have a very long lifetime due to a relatively low duty cycle, and the costs of energy from the combined systems would be dominated by the capital costs rather than by the fuel costs. Local and regional studies would be required to determine the optimal mix of such technologies.

The study also concludes, provisionally, that multi-bladed, wind-driven pumps, of the type once widely used in North America, would be competitive with diesel-powered irrigation for situations in which the mean annual wind-speed exceeds 3-4 m/s.

The solar options that have been discussed in relation to the developing countries are appropriate at certain levels of modernization and industrialization. The challenge in the near future, as solar options become available, may be to integrate the best of modern materials and technologies in a manner that is both simple and effective and that can be adapted to local needs, skills, and resources. It is understood, however, that when a country decides on technologies appropriate to a certain level of development, such decisions will be based on a recognition that its energy requirements will change as it develops, thus requiring more complex levels of technology.

GEOTHERMAL ENERGY

Geothermal energy is an indigenous resource in all countries on earth. Early uses included space heating and medicinal mineral baths. More recent are agricultural and industrial heating. Today it is sought primarily for the production of electric power, but space and industrial heating continue to be significant uses in many nations seeking to conserve fossil and nuclear fuels.

Attempts to generate electricity from geothermal energy date back almost to the turn of the century. The first experimental electric power generator was operated at Larderello, Italy, in 1904, capturing naturally flowing geothermal steam. Power has been produced there continuously since 1913. In a few other countries, including Japan, the United States, and New Zealand, attempts to develop geothermal energy for power production began in the early 1920s but were not commercially successful until the early 1960s. Geothermal energy developments have since been initiated in many countries, reaching a total electric capacity of about 1325 MW by 1976. Until recently the incentive for developing geothermal energy was low because fossil fuels seemed abundant and because energy planners traditionally were reluctant to depend on an energy resource whose quantity could not be precisely predicted or whose cost could not be known well in advance. Since 1973, when worldwide fossil fuel costs rose markedly, reexamination of all energy resources has brought about a renewed interest in geothermal energy for both electric and nonelectric uses.

The normal temperature gradient within the earth is about 25 ^{o}C/km depth. Thus, if the ambient temperature is 15 ^{o}C, it is expected that a random hole drilled to a depth of one kilometer will encounter a temperature of 40 ^{o}C. However, in some regions the temperature gradient is much greater than normal, as much as 1°C/m. Such gradients are usually associated with young volcanism, thin crust, or tectonic plate boundaries, and associated source temperatures may reach 1200°C.

Geothermal resources may be be divided into four basic types: liquid-dominated hydrothermal, vapor-dominated hydrothermal, petrothermal (hot dry rock), and geopressured. These types are distinguished by their thermodynamic and hydrologic characteristics.

Geothermal resources are called liquid-dominated if the pore space in the reservoir is filled with water (or a brine solution) that circulates by convection to transfer heat energy from the underlying rock matrix. Typical temperatures in such systems range from near-ambient to as high as 360°C. Exploration has revealed that of the hydrothermal resources near the earth's surface, liquid-dominated systems are far more abundant than vapor-dominated.

In the best liquid-dominated regions, temperature increases rapidly with depth but becomes almost constant in the main body of the reservoir because of convective flow and heat transfer. The reservoir fluid is saturated water at hydrostatic pressure.

Sometimes referred to as dry stream fields, vapor-dominated systems are relatively rare. However, the most economically favorable geothermal power developments in the world today are associated with vapor-dominated systems (at Larderello, Italy; The Geysers, California; and Matsukawa, Japan). In these systems, saturated steam is the continuous fluid phase in most of the pore space. Most investigators assume that water exists either as pore space droplets or in bulk volume at the bottom of the reservoir. Temperatures typically range from 220°C to 250°C. Production is relatively simple, and a slight superheating of steam often occurs during extraction.

It is reasonable to assume that far more heat is stored in the rock matrix than in circulating water. Since porosity generally decreases and temperature increases with depth, it is likely that vast volumes of hot dry rock exist at great depths within the earth's crust, in both solid and molten forms. Methods are being investigated for introducing cold water into hot dry rock systems (having natural or artificially induced fractures) and extracting heated water through wells adjacent to the injection holes.

Geopressured reservoirs are located in deep sedimentary strata where compaction has taken place over extended periods of time and where an effective shale seal has formed. Under these conditions, water is squeezed out of the shale matrix into adjacent sand bodies, becoming compressed well beyond the hydrostatic pressure at that depth. In some cases, geopressured water approaches lithostatic pressure. These systems are often characterized by higher-than-normal temperatures, as high as 237°C in some reservoirs along the Gulf Coast of the United States, with wellhead pressures of 76×10^6 Pa (11,000 psig) or more. Additionally, geopressured waters, or brine solutions, typically contain dissolved methane gas, probably to the point of saturation in many cases, and may become an economically important hydrocarbon resource in the future.

At present, the engineering and economic feasibility has not been established for either petrothermal (hot dry rock) or geopressured resources.

The geothermal resource could play an increasing role in overall energy supply. Some countries, fortunate to be in the major geothermal belts of the world, may be able to obtain significant fractions of their energy requirements from geothermal resources. Even countries underlain only by normal temperature gradients may find that geologic, climatologic, and economic circumstances combine to create a favorable economic setting for utilizing subterranean heat.

The purpose here is to review the prospects for geothermal energy over the next fifty years, summarizing its practical uses and its potential. It is hoped that this initial assessment will be useful in specific national investigations of how geothermal energy can play a role in meeting world energy requirements and that it can be updated in the future. Our present results are summarized in figures 1 and 2.

Prospects for geothermal energy utilization have been developed from two information sources. The first is a combination of the open literature and responses to a survey questionnaire directed to member nations of the World Energy Conference (and to nonmembers as well). The second information source is a resource base calculated from gross geologic and geophysical parameters. Worldwide geothermal resource base and energy utilization estimates,

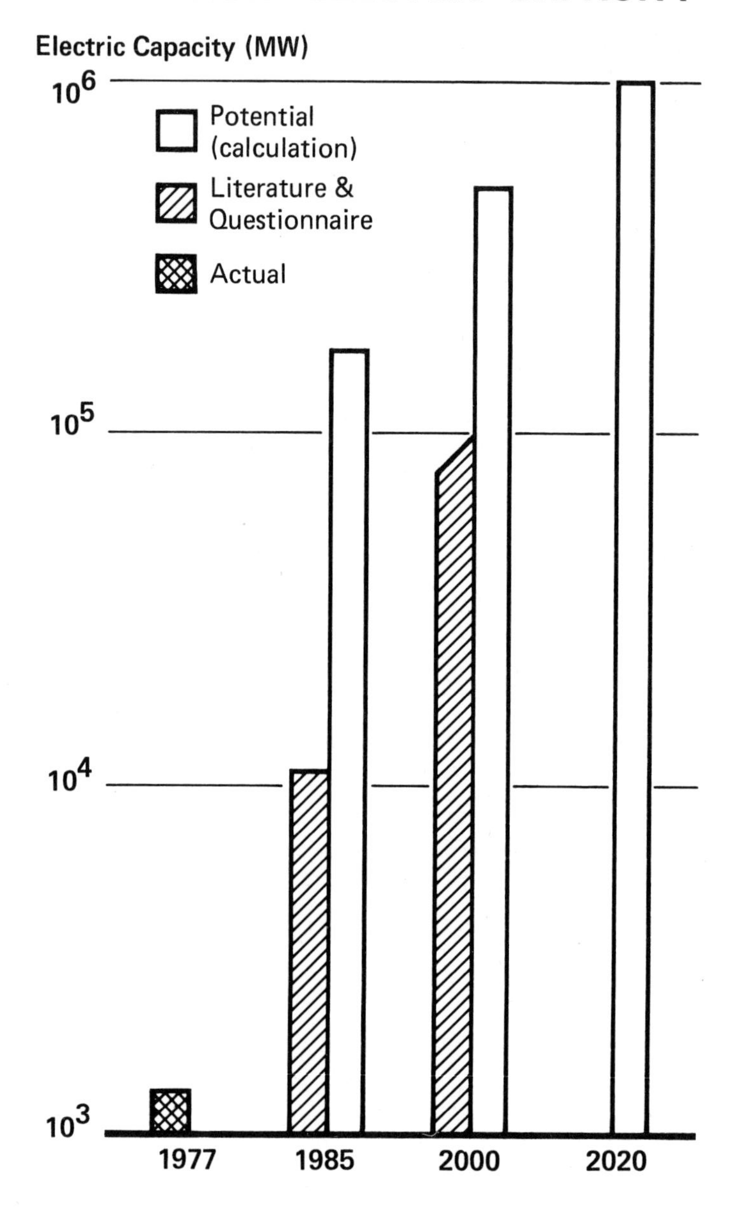
WORLDWIDE GEOTHERMAL ENERGY UTILIZATION: ELECTRIC CAPACITY
Electric Capacity (MW)
10^6
10^5
10^4
10^3
Potential (calculation)
Literature & Questionnaire
Actual
1977
1985
2000
2020

Figure 1

WORLDWIDE GEOTHERMAL ENERGY UTILIZATION: THERMAL CAPACITY

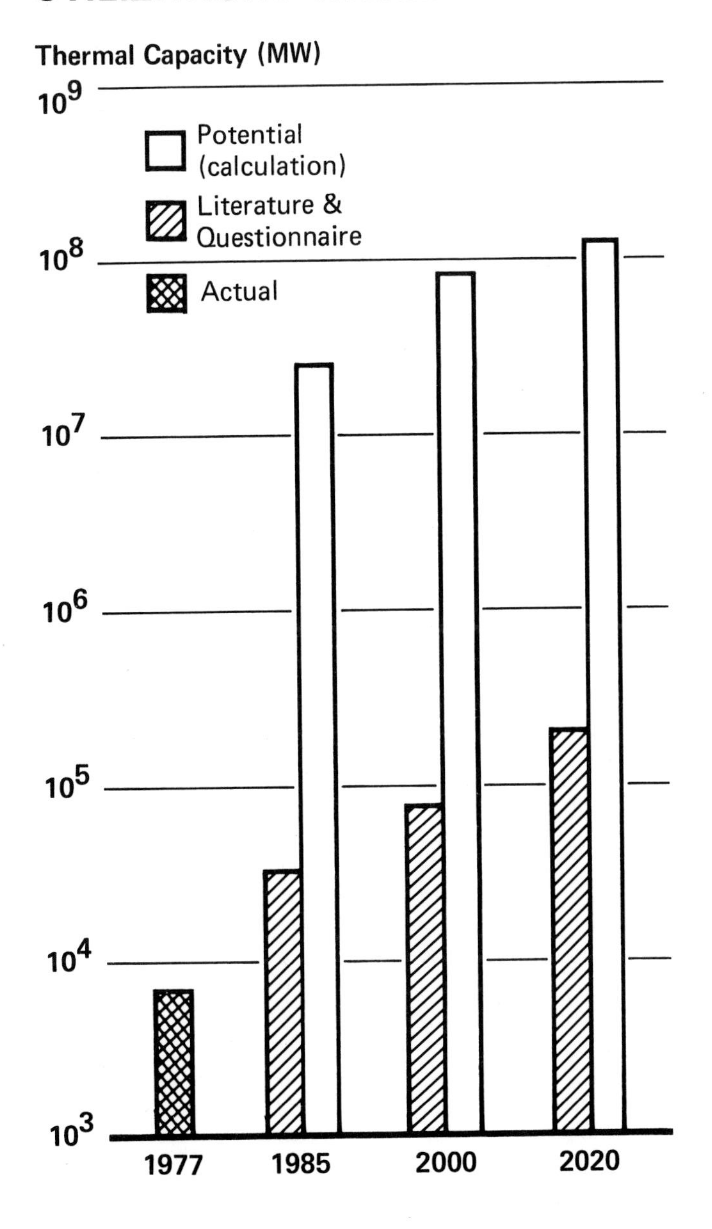

Figure 2

from both sources, are summarized in tables III and IV. (Note: In the tables, the notation E is used to represent the power of 10. For example, 2.1 E20 is equivalent to 2.1×10^{20}.)

Table III

WORLDWIDE GEOTHERMAL RESOURCE BASE

Source	Temperature (°C)	Resource Base (J)	Electric Potential (J)	Thermal Potential (J)
Literature and questionnaire	All	2.1 E24		
Calculation	<100	3.6 E25	0	2.6 E24
	100-150	3.8 E24	0	2.7 E23
	150-250	1.1 E24	1.7 E22	6.8 E22
	>250	7.3 E22	9.0 E20	3.5 E21
Total		4.1 E25	1.8 E22	2.9 E24

The sum of the resource base estimates reported by individual nations is 2.1×10^{24} J. The calculated base underlying the continental land masses (to a depth of 3 km and at temperatures above 15°C) is 4.1×10^{25} J. Only about 2% of this resource base is high enough in temperature to be considered for electric power production. Overall recovery and conversion efficiencies are also very small, yet the electric power potential is substantial: 1.8×10^{22} J, about 20% of which is thought to be producible by current technology. For perspective, this is the equivalent of 1.2×10^{6} MW of electric generating capacity operating for 100 years. These values assume that all areas having geothermal deposits suitable for electric power generation can be fully developed. It should be recognized that such complete development may not be practical or possible. The actual extent of development will depend on local conditions and regulations.

Table IV

WORLDWIDE GEOTHERMAL ENERGY UTILIZATION

Source	Electric Capacity (MW)			
	1977	1985	2000	2020
Literature and questionnaire	1.3 E3	1.1 E4-1.2 E4	7.6 E4-1.0 E5	Not available
Calculation		1.7 E5	5.0 E5	1.0 E6

Source	Thermal Capacity (MW)			
	1977	1985	2000	2020
Literature and questionnaire	7.0 E3	3.3 E4	7.6 E4	2.1 E5
Calculation		2.5 E7	8.0 E7	1.2 E8

In terms of total heat, the potential for nonelectric uses is far greater than for electric power generation; however, the expected growth of nonelectric applications is less than that for electric applications due to the fact that transportation of low-grade heat over any significant distance is usually uneconomical.

A regional breakdown of geothermal resource estimates obtained from questionnaire responses is given by Table V, while our corresponding calculated values, as explained in the main body of the report, are given by Table VI. The portion of this resource base which may be recovered and converted to electricity by existing technology has been estimated by respondents and is given in Table VII, while, once more, our calculated potential values are given in Table VIII. Finally, the estimates of nonelectric uses of geothermal energy as furnished by respondents is given by Table IX and should be compared with the value of our calculated global amount given in Table IV.

Examination of these tables will reveal that there exists a considerable discrepancy between the estimated values respondents to the questionnaire furnished and the considerably larger values we report as the recoverable electric or nonelectric potential. In part, this difference is due to the relatively little attention paid to geothermal exploration on a worldwide basis. Another factor is due to the relatively low priority afforded to geothermal energy utilization within many of the national energy plans of the world. Presumably, a more vigorous program of exploration and development could achieve targets approaching more closely the bounds provided by our calculations than would appear from the responses received from our questionnaire.

Table V

ESTIMATED GEOTHERMAL RESOURCE BASE

Region	Energy Content (J)
North America	1.7 E24
Central America	--
South America	5.4 E20
Western Europe	3.6 E23
Eastern Europe	5.4 E22
Asia	8.0 E19
Africa	--
Pacific Islands	1.3 E20
Total	2.1 E24

Source: Response to questionnaires and survey of literature. (Many countries either did not respond or did not have estimates.)

Table VI

CALCULATED GEOTHERMAL RESOURCE BASE
(joules)

Region	$<100^\circ C$	$100-150^\circ C$	$150-250^\circ C$	$>250^\circ C$	Total
North America	7.1 E24	1.2 E24	3.2 E23	2.0 E22	8.6 E24
Central America	1.5 E23	8.8 E22	2.8 E22	1.7 E21	2.7 E23
South America	4.6 E24	8.3 E23	2.3 E23	1.4 E21	5.7 E24
Western Europe	1.5 E24	7.1 E22	1.9 E22	1.2 E21	1.6 E24
Eastern Europe	6.7 E24	2.5 E23	6.3 E22	4.0 E21	7.0 E24
Asia	7.5 E24	7.9 E23	2.1 E23	1.3 E22	8.5 E24
Africa	5.2 E24	3.3 E23	7.0 E22	9.4 E21	5.6 E24
Pacific Islands	3.3 E24	2.6 E23	1.7 E23	1.0 E22	3.7 E24
Total	3.6 E25	3.8 E24	1.1 E24	7.3 E22	4.1 E25

Source: Methodology presented in body of report.

Table VII

ESTIMATED GEOTHERMAL ELECTRIC CAPACITY
(MW)

Region	1977 (installed)	1985	2000
North America	577	6410-7410	21 500-40 000
Central America	60	690-760	300-400
South America	--	20	--
Western Europe	424	800	1005
Eastern Europe	5.7	--	--
Asia	0.5	400	1000
Africa	--	30	60-90
Pacific Islands	258	2780-2850	52 100-57 600
Total	1325	11 130-12 475	75 965-99 890

Source: Response to questionnaires and survey of literature.

Table VIII

CALCULATED GEOTHERMAL ELECTRIC CAPACITY POTENTIAL

Region	Energy Potential E_b* (MWcen)	Generating Capacity (MW)		
		1985	2000	2020
North America	2.3 E5	35 000	106 000	212 000
Central America	1.7 E4	3 000	8 000	16 000
South America	2.6 E5	40 000	117 000	231 000
Western Europe	1.7 E4	3 000	8 000	15 000
Eastern Europe	7.2 E4	11 000	33 000	65 000
Asia	2.3 E5	34 000	100 000	200 000
Africa	9.0 E4	14 000	41 000	82 000
Pacific Islands	1.9 E5	30 000	90 000	179 000
Total	1.1 E6	170 000	503 000	1 000 000

Source: Methodology presented in body of report.

*Hydrothermal portion producible with current technology: 20% of total geothermal electric potential.

Table IX

ESTIMATED NONELECTRIC GEOTHERMAL APPLICATIONS
(J/d)*

Region	1985	2000	2020
North America	3.4 E14	1.3-2.9 E14	0.8-1.7 E16
Central America	1.7 E15	2.9 E15	5.0 E15
South America	--	--	--
Western Europe (inc. Iceland)	1.0 E14	7.9 E14	1.0 E15
Eastern Europe	1.5-2.6 E11	0.9-2.2 E14	1.7-4.6 E14
Asia	2.7 E14	--	--
Africa	--	--	--
Pacific Islands	3.4 E13	5.0 E13	8.3 E13
Total	2.5 E15	5.1-6.9 E15	1.4-2.4 E16

Source: Response to questionnaires and survey of literature.

*8.64×10^{10} J/d = 1 MW.

FUSION POWER

Although a number of complex scientific and technical problems must be solved before fusion power becomes a practical reality, the production of energy by the fusing or nuclear rearrangement of certain elements at the light end of the periodic table promises a number of very attractive features. The supply of fuel is virtually inexhaustible, widely distributed, and readily available at a small cost. Fuel costs should play virtually no role in determining the economic advantages of fusion power. Fusion offers the potential of an electric power supply system rather similar to the central generating units now in use by the industrialized nations and, in principle, should fit in well with already existing infrastructure. Additional advantages may result from the fact that radiation hazards associated with fusion are of a different nature and severity than those associated with conventional nuclear reactors or proposed fission breeders. In addition, fusion reactions promise a certain diversity in that they lend themselves to uses other than the production of electricity.

After two and a half decades of modest but determined efforts to understand the physical principles underlying nuclear fusion, interest in fusion power development has increased rapidly over the last nine years. Many nations (the United States, the USSR, Japan, and the Euratom members, among others) have established significant fusion programs, and today scientists and engineers from almost every country in the world are involved in some type of fusion research. This is indeed appropriate, as fusion power promises its benefits, either direct or indirect, to all nations and peoples.

Fusion can be considered to be a nuclear analogue of chemical combustion. Both types of reactions are accompanied by the release of energy, but in each case the "fuel" must be raised to an "ignition" temperature before energy release begins. However, for the fusion case, both the temperature required for ignition and the amount of energy released per event are greater by many orders of magnitude. In fact, the high temperatures required for fusion reactions, approximately 100×10^6 °C for the most easily "combustible" fuel combination, present some of the major problems associated with a practical reactor. First, means must be developed to heat the fuels, which at relevant temperatures are highly ionized gases, commonly termed plasmas. Next, means must be found to maintain the highly energetic fuel nuclei at a sufficiently high density and for a sufficiently long time for a significant number of reactions to take place before the plasma is dissipated. The amount of fusion power produced will be a function of the plasma temperature, the amount of plasma present, and the nature of the fuel mixture. In order to have a useful reactor, the amount of energy released during the thermonuclear burn must exceed by a significant factor the amount of energy that was required to initiate the reactions. This implies simultaneously achieving conditions of adequate plasma temperature, plasma density, and a slow enough rate of energy escape from the plasma; in other words, it is desired that a large fraction of the fuel be burned each time it is brought up to reaction temperatures.

One important milestone in the development of fusion power will be the demonstration of its scientific feasibility. Although several slightly different definitions of the term have been advanced, it is convenient to describe a scientific feasibility demonstration as one in which more energy is produced by fusion reactions than was expended in causing them to occur. Since this definition implies energy recycling, it is apparent that improvements in energy conversion, as well as in plasma control, will contribute to such a demonstration.

The requirement for maintaining plasma density sufficiently long to permit an adequate number of fusion reactions can be approached in two different ways. The first, the more conventional approach, is to use any of a variety of magnetic field configurations to isolate and insulate the hot plasma core from its surroundings. The other approach utilizes the concept of inertial confinement where some energetic source, such as a high-power laser or intense beams of electrons or ions, is used to deposit large amounts of energy over a brief interval of time on a pellet containing fusion fuels to initiate the thermonuclear burn. This concept has been utilized in large explosive thermonuclear devices, of course, but the phenomenon must be demonstrated on a greatly reduced scale before it will be suitable for peaceful commercial applications.

It might be noted that plasma containment on the sun, where a chain of fusion reactions ultimately transforms hydrogen nuclei to helium nuclei, positrons, and neutrinos, is achieved by that body's immense gravitation field. However, the fusion processes in the sun proceed at a very slow rate; and, in any case, this method of containment is, of course, not feasible for terrestrial reactors.

Although the problems associated with the fusion process have proved to be considerably more difficult than was originally envisioned some 25 years ago, when the quest for harnessing fusion power began in earnest, progress in understanding the behavior of these plasmas has been impressive and highly encouraging over the past decade. Because our knowledge of the underlying physics, extensive as it is, is not perfect, one cannot predict with certainty when, or even if, scientific feasibility will be achieved. However, recent achievements in plasma production, plasma heating, and plasma confinement (the art of preventing too rapid a loss of energy from the plasma) have led to a widespread belief among the scientific community that this goal will be reached within the next 5-10 years by one or more of several approaches under current investigation.

As our confidence in the scientific feasibility of fusion increases, it becomes desirable to turn increasing attention to the engineering problems that must be solved before fusion power can become a practical reality. The challenges presented here include the development of suitable materials to withstand the hostile environment of a fusion core, of appropriate blanket designs, of systems to preclude any potential hazards to human health or the environment, and of engineering components necessary for plasma heating and confinement and energy conversion. These problems, collectively, may present a challenge equal to, or even greater than, those involved in solving the basic physics problems associated with fusion. Finally, it will be necessary to design systems that are not only consistent with the limits to the state of the art achieved but also compatible with and attractive to the utility system for which the fusion power plant is destined.

There are a number of fusion fuel cycles using light elements, such as the isotopes of hydrogen, helium (He), lithium (Li), and boron (B) that might be used to produce energy. The fuel cycle that utilizes two isotopes of hydrogen, deuterium (D) and tritium (T), has the lowest requirements on plasma core heating and confinement, has the highest probability of occurrence of all the possible fuel cycles, and releases a significant amount of energy. For these reasons the bulk of fusion research to date has been concentrated on this fuel cycle, and it will almost certainly be used for the scientific feasibility demonstration and probably for the first-generation power reactors.

The supply of deuterium, which can be easily separated from water, is virtually inexhaustible. The waters of the earth contain approximately 46×10^{12} t of deuterium, and burning this amount of fuel to completion would produce 16×10^{30} J of thermal energy.

On the other hand, tritium, which has a relatively brief radioactive lifetime, must be created artificially. This can be done relatively straightforwardly since the fast neutrons resulting from the fusion reaction can be used, in part, to breed tritium from lithium. It may appear that breeding would be difficult since only one neutron is produced per fusion event. In fact, however, the fusion neutron can readily multiply in a suitably designed blanket, and a number of preliminary nuclear designs indicate that the question of neutron economy does not threaten to be a major stumbling block on the way to fusion reactor developments. The amount of lithium consumed by breeding tritium can vary with reactor design. A representative number, however, is 1.8 kg of natural lithium for each MW (electric)-yr of electricity generated, assuming a thermal efficiency of 40%. Current estimates of the world's recoverable lithium reserves and resources are about 10×10^{6} t, exclusive of the much larger abundance of dissolved lithium in the oceans. It seems rather unlikely that lithium availability could restrict the use of fusion to supply the world's electricity needs for many hundreds and, possibly, thousands of years. In effect, then, the D-T-Li fusion breeder would use deuterium (which occurs in the proportion of 1:6700 with ordinary hydrogen and is in virtually unlimited supply) and lithium (which, while not nearly as abundant as deuterium, is still quite plentiful).

In addition to the problems associated with producing tritium for fuel, reactors utilizing the D-T reaction will face special difficulties that must be resolved if the engineering feasibility of fusion is to be realized. The problems are associated with tritium handling, first-wall thermal loading, heat removal, and the formidable difficulties of material integrity. Problems of accessibility and maintainability of the activated structural material are also formidable but may be minimized if high-performance, low-activation structural material can be developed. Offsetting these disadvantages of the D-T fuel cycle is the fact that the reaction can provide a rich source of neutrons. These could be used not only to produce tritium but also to breed fissile atoms in special blankets containing fertile nuclear fuels, such as uranium or thorium. This offers the possibility of extending the fuel supply for fission reactors. These fusion-fission systems combine some of the features of both fission and fusion.

Using the deuterium-deuterium (D-D) reaction eliminates the requirement for tritium production. This fuel cycle simplifies blanket design problems but puts more demanding limits on the temperature, the plasma core density, and the energy confinement time.

Other fusion fuel cycles, such as D-^{3}He, p-^{6}Li, D-^{6}Li, p-^{7}Li, or p-^{11}B, present even greater physics problems than the D-D reaction; however, they share a common interesting feature in that most of the fusion energy is released in the form of charged-particle kinetic energy, minimizing or even eliminating neutron production. This feature opens the possibility, in principle, of very high efficiency energy conversion. These neutronless fuel cycles are now being evaluated to see if the problems associated with tritium handling and the first-wall and blanket radiation can be reduced. Accessibility and maintainability could be improved, thermal pollution problems minimized, and safety features simplified. Although low power density and high circulatory power are characteristics of these systems, the prospect for using these cycles improves as the ability to heat and confine fusion plasma advances and as efficient means of converting and utilizing radiant energy are developed.

Although the major research efforts in the fusion program throughout the world have been focused on large, D-T based reactors designed to produce electric power, investigation has begun into alternative systems and uses. For example, the large neutron and other radiation fields possible with fusion reactors could be used for the production of fissile material for fueling conventional nuclear reactors, for the production of chemical fuels, or for the transmutation of radioactive wastes from fission reactors. Research is also being conducted into the possibility of developing a series of reactors of different sizes, configurations, and operational characteristics. This potential versatility of application could be an important asset of fusion power.

Based on the above considerations, we conclude that the potential contributions of fusion are great and could even be commanding in the course of time. A variety of the approaches now under active consideration could succeed, or other approaches not yet well explored could prove to be the ultimate favorites. The state of the art is embryonic by comparison with many other alternative energy conversion technologies, and it would be premature to say with any conviction when fusion might begin to enter the commercial markets in a serious way. If present indications of the rate of future progress hold firm and the world's fusion research and development effort continues to be vigorously supported, fusion power could well become a contender in the marketplace during the early decades of the next century.

ENERGY CONSERVATION

Lasse Nevanlinna
System Planning Department, Imatran Voima Osakeyhtiö, Finland

Fjalar Kommonen
EKONO Oy, Finland

Units

Units, their abbreviations and conversion factors used are:

- Million tonnes or kilograms of oil equivalent = Mtoe, kg oe
  (used for fuels and total energy)
- ExaJoule = 10^{18} Joule = EJ
  (used for total energy)
- Terawatthours (= 10^{12} Wh), kilowatthours = TWh, kWh
  (for electrical energy)

1 Mtoe = 0.044 EJ

1 TWh = 0.1...0.28 Mtoe [1)]

1 kWh = 0.1...0.28 kg oe [1)]

---

1) Conversion factor varying, and specified in the text.

ENERGY CONSERVATION

Contents

1. Introduction

2. Definition of conservation and its relation to Gross World Product GWP

3. Presentation of results
   3.1 Transport
   3.2 Industry
   3.3 Domestic/public/commercial
   3.4 Total consumption
   3.5 Energy sector own use
   3.6 Electricity production
   3.7 Total energy use
   3.8 Timing of conservation

4. Conclusions

References

Appendix 1

Appendix 2

Abstract

The objective of this study was to assess the global significance of energy conservation until year 2020. Starting from the fact that the ratio of total energy demand (TED) to gross world product (GWP) has been nearly constant in the past, the possibilities of a development towards a lower TED/GWP ratio have been analysed. The study concludes that with rising energy prices and a world wide public concern about the depletion of conventional energy resources as incentives, significant conservation possibilities exist in all consumption sectors. Together with structural changes in the world energy demand these possibilities could utilized result in a TED/GWP ratio, which in 2020 is only nearly half the present value. The electricity demand to GWP ratio would remain about at the present level, but the electricity production offers a considerable fuel conservation potential. The influence of nuclear electrical energy price in relation to fuel prices upon energy conservation solutions is essential and needs much attention in future technology development.

A relatively high global economic growth rate of 4.3 % p.a. was chosen for the projection, but the result is believed to be rather insensitive to variations in economic growth. It is also concluded that conservation offers a favourable alternative "energy source" for the transition period at the end of this century, when a continued oil production increase seems not possible.

1. Introduction

The objective of this study is to estimate the global energy conservation potential until 2020 by main consumption sector. The study is based on the present state of technology and its estimated development and includes the influence of policy actions and conservation measures already initiated in many countries (1)(2)(3). The extent of possible conservation has not been evaluated as a function of energy price. Rather, the conservation measures applied concentrate on oil conservation and will be economically feasible at real consumer energy price level of today or slightly higher, provided that their technological development has reached the required maturity.

The total energy demand has been divided in five main sectors:

- Transport
- Industry
- Domestic/public/commercial
- Energy sector own use
- Electricity production

In evaluating the total energy demand and conservation the demand for electricity and other energy demand ("fuel") have been treated separately in each sector. Then the primary energy demand of electricity production has been treated as one independent sector. This approach is in our opinion important, because the demand for electrical energy has grown, and will according to the programs of many countries even continue to grow considerably faster than other energy demand. As electricity generation by means of the condensing process has a relatively low thermodynamic efficiency, this sector is already now a major consumer of primary energy. On the other hand it shows interesting possibilities for conservation by adopting combined production of heat and electricity ("cogeneration"), and by improving the thermodynamic efficiency of the generation process. Also in view of different policies for future energy supply, e.g. the role of nuclear energy, it is essential to know how the total energy demand is split between electricity and other energy forms.

As base data the global energy balance and its sectorial breakdown for the year 1975 given in Table 1 has been used.

TABLE 1

Year: 1975 Region: World

Energy demand

Sector	Primary energy input 2)			Electricity		
	Mtoe	kg oe p.c.	%	TWh	kWh p.c.	%
1. Transport	1000	256	16·6	100	26	1·8
2. Industry	1500	385	24·8	2700	692	47·4
3. Domestic, public, commercial	1300	333	21·5	2300	590	40·4
4. Total consumption	3800	974	62·9	5100	1308	89·5
5. Energy Sector own use	400	103	6·6	600	152	10·5
6. Electricity Production (net) 1)	1600	410	26·5	5700	1460	100
7. Total energy use	5800	1487	96·0			
8. Feedstocks, etc.	240	62	4·0			
9. Total use	6040	1550	100			

Population: $3{\cdot}9 \times 10^9$

Division in Sources:

	Total Use		Electricity Production	
	Mtoe	%	Mtoe	%
Oil	2500	41·4	350	21·9
Gas	1040	17·2	200	12·5
Coal	1500	24·8	650	40·6
Wood etc.	600	9·9	-	-
Nuclear 1)	50	0·8	50	3·1
Hydro 1)	350	5·8	350	21·8
Solar	-	-	-	-
Total	6040	100	1600	100

1) 0·28 kg oe/kWh

2) Items 1...5 excluding electrical energy

2. Definition of conservation and its relation to the Gross World Product (GWP).

Defining long term conservation offers certain difficulties. As a reference against which the impact of conservation is measured we have chosen the ratio between Total Energy Demand (TED) and Gross World Product (GWP), which is known to have remained nearly constant in the past.

The goal of energy conservation is to achieve an appropriate economic growth with minimum total energy consumption.

This will not necessarily mean that the ratio between energy consumption and GNP has to decrease in all sectors under all circumstances. In countries, where electrification is in an early stage of development, electricity demand obviously will grow faster than GNP to obtain an appropriate economy growth. The same is valid for a country which wishes to make the transition from an oil based energy economy to a nuclear based economy in a relatively short period of time. But in general the goal of conservation is to decrease the ratio TED/GWP.

For the purpose of this study a GWP growth in the periods 1975-2000 and 2000-2020 of 4.6 % and 4.1 % p.a. respectively is assumed. Further it is assumed, that the developing countries grow faster than the average, around 6 to 5 % p.a., and the developed countries correspondingly slower or 4 to 3 % p.a. This has a certain influence upon the energy growth and conservation pattern. For example the electricity demand will grow faster than with an even GNP growth distribution. On the other hand lead times for global conservation may be shorter, provided, that conservation action is vigorous in the developing world.

With these assumptions the GWP will develop as follows:

Year	1975	1985	2000	2020
GWP, 10^9 \$ (1972 value)	4400	6900	13500	30000
GWP ratio	1.0	1.57	3.07	6.8

To evaluate the conservation potential the energy demand by sector is compared for three cases, namely:

i) Constant ratio or reference case, where the energy demand to GWP ratio remains constant

ii) Structural conservation only case, where only the impact of structural change in GWP and energy demand have been accounted for. Specific energy use for industrial production (energy input per unit of product) for example is unchanged, but the share of energy-intensive production may be lower in the total industry production mix. Other examples of structural change are that the volume of heated building space or the number of automobiles may grow at a slower rate than GWP.

iii) Maximum conservation case, where all measures have been included that appear to be applicable until 2020.

3. Presentation of results

3.1 Transport

The energy conservation potential in the transport sector is considerable but difficult to estimate, because it is heavily depending upon government policies in the areas of fuel and vehicle taxation, and upon the ability of public transportation to compete with private cars, which can be considered a "life style" issue. For the present almost 70 % of total energy demand in the transport sector (1000 Mtoe per annum) is consumed in the OECD countries, and 90 % of it is depending on oil. The share of road traffic is almost 60 %.

Our estimate is based upon two assumptions:

- the number of cars grows slower than GWP (ratio assumed to be 1:1.7)
- the automobile fuel consumption will be 60 % of the present.

These assumptions lead to an energy demand in the road transport sector which is over 50 % smaller per GWP unit than the present. On the other hand the international transport volume by air, ship, and railroad will grow faster than the GWP. Assuming that this volume increase is compensated by lower specific fuel consumption, the overall fuel conservation potential of this sector is estimated to 40 %.

The electricity consumption in transportation will grow faster than GWP because of vigorous development of electrified rail transport and urban transit systems including electrical vehicles.

Summary of transport energy demand:

	Fuel		Electricity	
	Mtoe	ED ratio[1)]	TWh	ED ratio[1)]
1975	1000	1.00	100	1.00
2020 "const. ratio"	6800	1.00	700	1.00
"struct. conservation"	5100	0.75	1000	1.44
"max. conservation"	4100	0.60	1000	1.44

3.2 Industry

3.2.1 Fuel

The fuel demand per GWP unit has in the industry sector decreased for quite a long time. The main reasons for this are that the share of industry in the total gross products has decreased, and that the structure of industrial production is changing towards less energy consuming branches in the consumer goods sector. Although this trend will not be evident in all developing countries during the time period we are considering, it is believed, that the world average will follow the previous trend. This continuing structural change will reduce the fuel consumption per GWP unit by 10 % from 1975 to 2020.

---

1) ED ratio = sector energy demand/GWP. 1975 value = 1.0

Although the industry sector is the most energy cost-conscious sector, and reacts well at energy price changes when building new plants, energy conservation investments in existing installations as a rule are given rather low priority, even in the most energy-intensive branches like steel or pulp and paper. A solution here is government guaranteed loans for conservation investments.

The possibilities for long term energy conservation in industrial processes have been estimated based upon extensive material available for what in this report must be defined as short and medium term (4)(5)(6). In view of these results, of theoretical improvement potential (7), and including the anticipated result of research and development applied, we assume the fuel conservation possibility to be in the range of 30 to 50 % in this sector.

What actually will be achieved is also depending of how the price ratio of nuclear energy to fossile fuel is developing. If nuclear electricity becomes lower priced, industrial processes will increase their electricity consumption and decrease fuel consumption.

In view of the aspects presented our conservative estimation of fuel conservation in industry is 40 %.

3.2.2 Electricity

The electricity demand per GWP unit in industry has in the past shown a steady growth although slowing down in recent years. Even with a saturation in electricity demand this ratio would in 2020 be at least 1.1 times the present value. It is assumed, that conservation efforts will bring down the electricity demand per GWP to remain constant.

3.2.3 Total

The total energy demand of the industrial sector then becomes as follows:

	Fuel		Electricity	
	Mtoe	ED ratio	TWh	ED ratio
1975	1500	1.00	2700	1.00
2020 "constant ratio"	10200	1.00	18400	1.00
"struct. conservation"	9000	0.89	20000	1.09
"max. conservation"	6000	0.59	18000	0.98

3.3 Domestic, public, commercial

3.3.1 Fuel

The fuel consumption in this sector is mainly due to energy demand for space and warm water heating in the northern parts of the world, representing developed countries with an energy consumption and living standard above average. Consequently the GNP growth in these countries is slower than the world average. The hot countries representing fast economic growth have little need for heating energy, which is likely to be covered largely by passive and active utilization of solar energy in the future.

In the northern countries the growth of heated building space will be below GNP growth, probably not exceeding 2 % per annum corresponding to a doubling time of 35 years. Based on these assumptions the fuel demand of this sector would without any conservation measures grow only half the amount of GWP growth until 2020, resulting in a fuel demand of 4400 Mtoe.

Space heating offers, however, a considerable conservation potential. Among different possibilities is mentioned the following:

- double pane windows in mild climate countries and triple pane windows in cold climate countries - energy saving 10 %
- improved wall insulation - energy saving 10 %
- improved control of heating and ventilation - energy saving 10 %.

The cumulative result of these three measures is 20-25 %. The implementation will take place mainly in new or renovated residential buildings, and has long lead times.

In public, commercial, and industrial buildings also recovery of heat from exhaust air will be considered. In cases where the heat generation from equipment and people is considerable, the external heat requirement may be negligible.

The application of heat pumps and utilization of solar energy represent advanced technology with correspondingly meager experience. Preliminary results indicate, that external heating requirements of a building could be reduced to 10-40 % by such techniques, in comparison to present levels.

With decreasing external requirements for heating energy, there is an increasing advantage from electrical heating. The fuel consumption for heating will in such case occur in the power plant, and is there slightly larger than in individually fuel-heated houses. The significance of this is small when the heat consumption of the house has been reduced by conservation measures. In certain cases an electrical heating system can show equal or better overall efficiency. Individually heated houses use normally oil or gas as fuel, whereas electricity production will mostly be based on coal or nuclear fuel. Thus by adopting electrical space heating, a substitution of oil and gas will take place.

Electrical heating is suitable for residential houses and certain commercial and public buildings. In high-density urban areas district heating will be applied where the heating season is appropriately long. The better efficiency of a centralized system then will yield a 3-10 % saving. The main advantages of a district heating system are, however, that oil can be substituted by other primary energy sources like coal or nuclear energy, and that cogeneration of heat and power is possible, reducing the energy consumption of electricity generation with more than half.

3.3.2 Electricity

Electricity demand of this sector consists mainly of appliances' consumption (lighting, kitchen equipment, office machines, refrigeration equipment, small pumps, electronics etc.). This kind of electricity demand per capita correlates strongly with GNP per capita. In year 2020 GWP per capita is assumed to 3400 $ per capita according to our base data. The corresponding electricity consumption according to experience in this sector is 1700 kWh p.c. or 15 000 TWh globally. The conservation potential in this sector is not very substantial. We have estimated it to 10 % due to lower appliances' consumption.

This estimate must be adjusted for an increased implementation of electrical heating of buldings. In 3.3.1 it was assumed a total fuel consumption of this sector (without conservation and neglecting electrical heating) of 4400 Mtoe in 2020. If the share of electrical space heating is increased to 20 %, it substitutes 900 Mtoe of fuel. The corresponding consumption of electrical energy is assumed to 3 kWh/kg oe (implementation in buildings with advanced conservation standard including heat pumps and solar energy utilization) or about 2700 TWh.

Air conditioning and cooling is a substantial energy consumer in warm climate countries. In the absence of base data on this important area we have assumed, that such energy demand partly is met by increased utilization of solar energy and partly is included in the above used GNP per capita correlation.

3.3.3 Total

To summarize we start from the fuel consumption of the sector which was estimated to 4400 Mtoe without conservation measures. As an oil conserving measure 900 Mtoe of this is substituted by 2700 TWh electrical heating, resulting in 3500 Mtoe. By the conservation measures mentioned in 3.3.1 40 % of this can be saved, leaving 2100 Mtoe as final fuel demand.

The electricity demand 15000 TWh is reduced by 10 % conservation = 1500 TWh and increased by 2700 TWh of electrical heating:

	Fuel		Electricity	
	Mtoe	ED ratio	TWh	ED ratio
1975	1300	1.00	2300	1.00
2020 "constant ratio"	8800	1.00	15700	1.00
"struct. conservation"	4400	0.50	15000	0.95
"max. conservation"	2100	0.24	16200	1.03

3.4 Total consumption - sectors 1 to 3

The sum of the previous three tables gives the following:

	Fuel		Electricity	
	Mtoe	ED ratio	TWh	ED ratio
1975	3800	1.00	5100	1.00
2020 "constant ratio"	25800	1.00	34800	1.00
"struct. conservation"	18500	0.72	36000	1.03
"max. conservation"	12200	0.47	35200	1.01

3.5 Energy sector own use

The energy demand of this sector is dependent upon total consumption discussed in item 3.4 above. The fuel demand is chiefly composed of oil refining and synthetic oil and gas producing plant fuel demand. (Fuel for electricity generation is not included here, but appears in item 3.6). Because of increased production of synthetic fuels from coal we have assumed the fuel demand of this sector to grow from 10.5 % in 1975 to 15 % in 2020 of total fuel consumption.

Electricity demand of this sector consists of transmission losses. (Power station own use is included in item 3.6 below). We estimate that improved transmission technology reduces these losses from 12 % in 1975 to 10 % in 2020:

	Fuel		Electricity	
	Mtoe	ED ratio	TWh	ED ratio
1975	400	1.00	600	1.00
2020 "constant ratio"	2700	1.00	4100	1.00
"struct. conservation"	2800	1.04	4300	1.05
"max. conservation"	1800	0.67	3500	0.85

3.6 Electricity production

The average fuel consumption of electricity production was in 1975 0.28 kg oe/kWh sent out. The amount of electricity from cogeneration of heat and power was negligible in global perspective. Today electricity is mainly produced in the steam turbine process, where the admission temperature to the turbine is 300-550 ^{o}C giving a 30 to 40 % efficiency. In the future the starting temperature of the thermodynamic process will increase to 900 ^{o}C by use of combined gas turbine-steam turbine process (fig. 1). The resulting efficiency increase is 10 %. Probably the MHD-process has by 2020 to some extent found use as an initial stage for the conventional steam process. We have therefore assumed an average fuel consumption in 2020 of 0.21 kg oe/kWh for condensing power generation.

The fuel demand for electricity production can also be reduced by cogeneration in cases, where a sufficiently large and steady heat load is available. Such loads can be found in energy-intensive industries and in district heating of urban areas. We have assumed, that 30 % of industry electricity demand can be met by cogeneration, and that 10 % of other electricity demand is produced in such processes, with a fuel consumption of 0.1 kg oe/kWh.

As a third possibility, although of smaller significance, to reduce fuel consumption of power generation, is the efficient recovery of waste fuel and waste heat for power generation, mainly in the petroleum refining, metallurgical, and chemical industry. We have here conservatively assumed, that such by-product power generation results in 170 TWh of electricity.

FIGURE 1

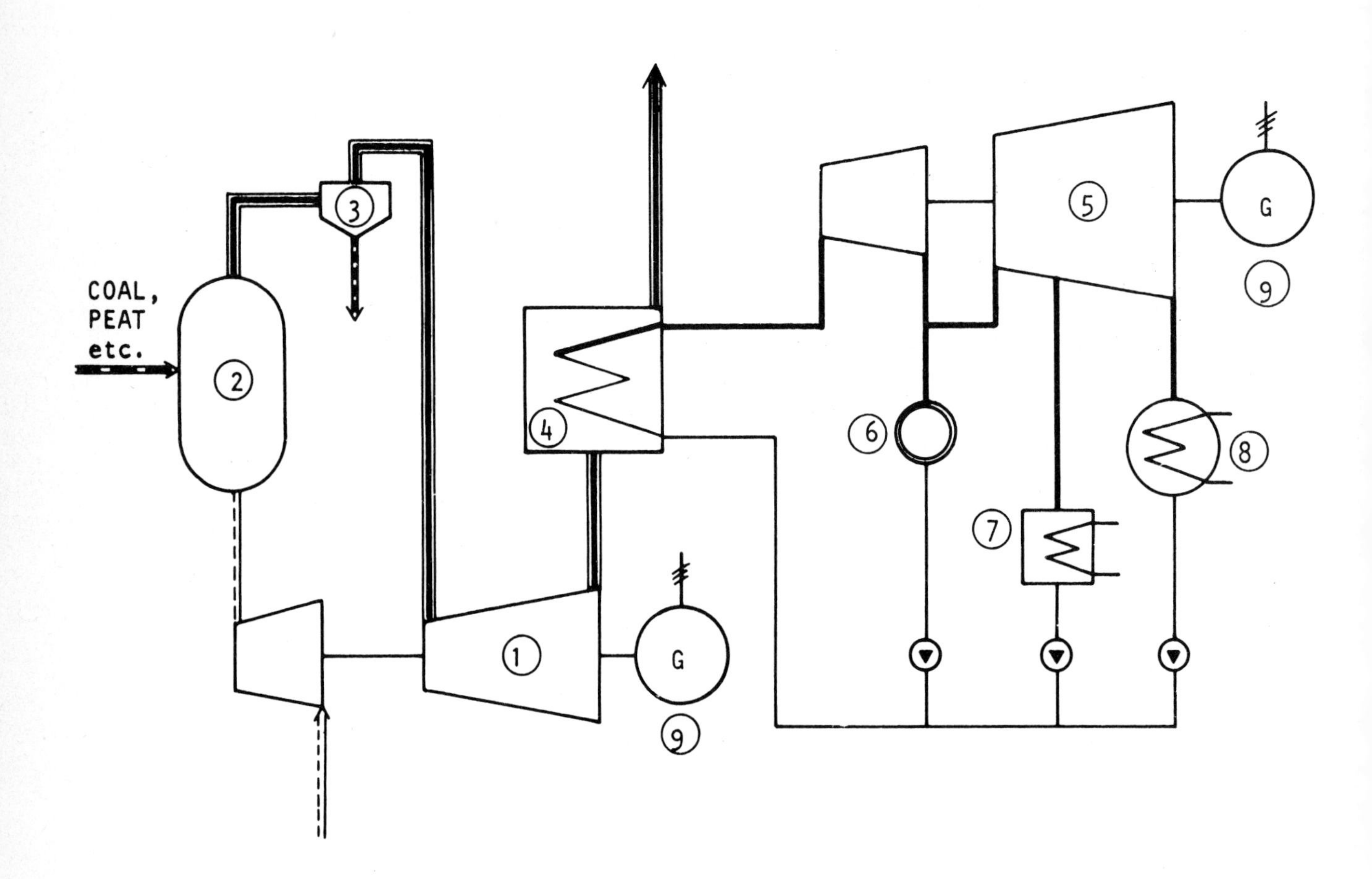

LEGEND:

1 GAS TURBINE
2 PRESSURIZED BURNER (FLUIDIZED BED)
3 GAS PURIFICATION (CYCLONS, GRANULAR)
4 WASTE HEAT BOILER
5 STEAM TURBINE
6 PROCESS STEAM USER
7 HEAT EXCHANGER GENERATING HOT WATER FOR DISTRICT HEATING OR PROCESS USE
8 CONDENSER
9 GENERATOR

As summary, the result for this sector is the following:

	Fuel		Electricity		
	Mtoe	ED ratio	TWh	ED ratio	kg oe / kWh
1975	1600	1.00	5700	1.00	0.28
2020 "constant ratio"	10900	1.00	38900	1.00	0.28
"struct. conservation"	11300	1.03	40300	1.03	0.28
"max. conservation"	7300	0.67	38700	1.00	0.19

3.7 Total energy use

The total energy use in 2020 for all three cases is presented in detail in Table 2, and is summarized below:

Case	i "constant ratio"		ii "structural conservation"		iii "maximum conservation"	
	Mtoe	ED ratio	Mtoe	ED ratio	Mtoe	ED ratio
1. Transport	6800	1.00	5100	0.75	4100	0.60
2. Industry	10200	1.00	9000	0.88	6000	0.59
3. Dom./publ./comm.	8800	1.00	4400	0.50	2100	0.24
4. En. sector own use	2700	1.00	2800	1.04	1800	0.67
5. Electr. production	10900	1.00	11300	1.04	7300	0.67
6. Total energy use	39400	1.00	32600	0.83	21300	0.54

It is seen that the total primary energy demand in 2020 with maximum conservation is estimated to be 46 % lower than the "constant ratio" case implies, and that of this 17 % is estimated to be due to structural change.

The conservation potential can now be calculated as difference between the various cases. Below is given an analysis of how the sectors contribute to conservation potential:

TABLE 2

Year: 2020 Region: World

Sector	"Constant ratio" case				"Structural conservation" case				"Maximum conservation" case			
	Fuel Mtoe	%	Electricity TWh	%	Fuel Mtoe	%	Electricity TWh	%	Fuel Mtoe	%	Electricity TWh	%
1. Transport	6800	17.3	700	1.8	5100	15.6	1000	2.5	4100	19.2	1000	2.6
2. Industry	10200	25.9	18400	47.3	9000	27.6	20000	49.6	6000	28.2	18000	46.5
3. Domestic, public, commercial	8800	22.3	15700	40.4	4400	13.5	15000	37.2	2100	9.9	16200	41.9
4. Total consumption	25800	65.5	34800	89.5	18500	56.7	36000	89.3	12200	57.3	35200	91.0
5. En. sector own use	2700	6.8	4100	10.5	2800	8.6	4300	10.7	1800	8.5	3500	9.0
6. Electricity production	10900	27.7	38900	100.0	11300	34.7	40300	100.0	7300	34.2	38700	100.0
7. Total energy use	39400	100			32600	100			21300	100		

	Total potential[1]		Structural change[1] excluded	
	Mtoe	%	Mtoe	%
1. Transport	2700	15	1700	14
2. Industry	4200	23	3000	25
3. Dom./public/commercial	6700	37	2300	19
4. En. sector own use	900	5	1100	9
5. Electricity prod.	3600	20	4000	33
6. Total	18100	100	12100	100

In the above table the conservation potential is given in terms of fuel. Alternatively the sectors 4 and 5 can be divided upon the final consumption sectors 1 to 3 giving the total energy conservation potential for these sectors:

	Total potential		Structural change excluded	
	Mtoe	%	Mtoe	%
1. Transport	2900	16	2200	18
2. Industry	6500	36	6200	51
3. Dom./public/commercial	8700	48	3700	31
4. Total	18100	100	12100	100

1) Total potential = case i - case iii; Structural change excluded = case ii - case iii

3.8 Timing of energy conservation

In the preceeding text we have defined the overall conservation potential, which is achievable with today known technology within some 40 years. In reality energy conservation is a function both of economical growth and of time. For todays decisions it is essential to know how fast the conservation can be implemented. In general substantial conservation will have considerable lead times because energy using and converting equipment and buildings cannot be replaced very fast. The development of energy saving technology will also require time.

With respect to different consumption sectors the following can be said:

- The transport sector offers the fastest possibilities, as the automobile fleet, where the conservation can be substantial, has an approximate turnover time of 10 years. Important however is that the conservation efforts gain a rapid approval from the car users and manufacturers.

- In industry and in energy conversion plants part of the conservation can be implemented very fast by attachments to and improvement of present plants, amounting to 5-10 %. The larger part of conservation can only be achieved in the course of the erection of new plants, and require process development. The lead times here are about 10 to 20 years.

- In domestic, public, and commercial sector the longest lead times can be found in the case of building energy demand. The life time of buildings is long and new buildings represent thus a comparably small part. When new methods have gained general acceptance, the conservation impact will be strong towards the end of the century.

The estimated interrelations between conservation potential, GWP, and time are presented in figure 2. The analysis presented in summary above results in a progressive growth of the conservation impact up to 2000, when the growth of conservation reaches its maximum. The total impact of all conservation measures is then largest. After 2000 the conservation impact slows down, as the potential by then is nearly fully utilized.

4. Conclusions

The following table summarizes our findings on conservation and its timing 1975 to 2020:

Year		1975	1985	2000	2020
GWP, 10^9 \$ (1972 value)		4400	6900	13500	30000
GWP ratio (1975=1)		1.0	1.6	3.1	6.8
Total Energy Demand TED,					
TED ratio (TED/GWP; 1975=1.0)					
i) "Constant ratio"	Mtoe	5800	9100	17800	39400
	ExaJoule	255	400	780	1730
	TED ratio	1.00	1.00	1.00	1.00
ii) "Structural conservation only"	Mtoe	5800	8600	15800	32600
	ExaJoule	255	380	700	1430
	TED ratio	1.00	0.94	0.89	0.83
iii) "Maximum conservation"	Mtoe	5800	8300	11900	21300
	ExaJoule	255	360	520	930
	TED ratio	1.00	0.91	0.67	0.54

The above result is believed to be rather insensitive to variations in economic growth. Thus the main conclusion of this study is condensed in the TED/GWP ratio figures above and the corresponding curves in the upper part of Figure 2.

FIGURE 2

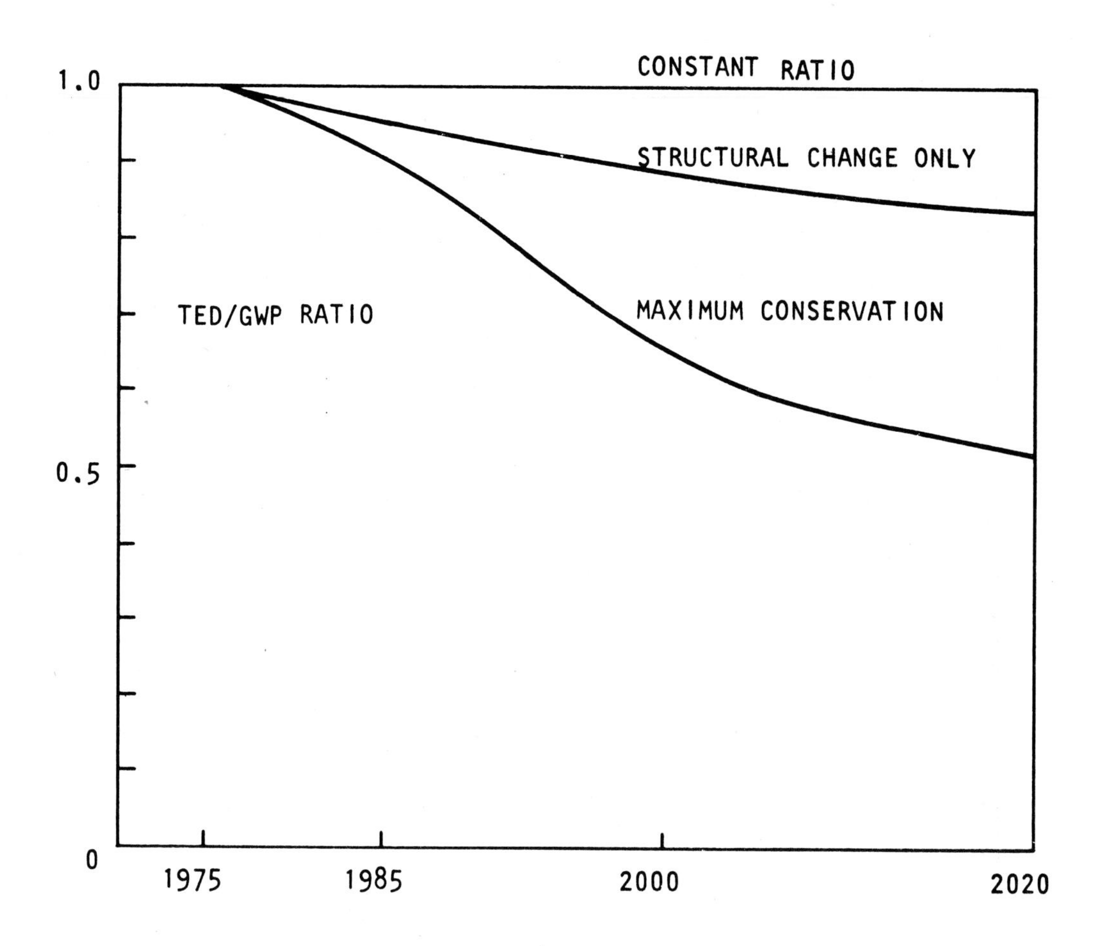

1.0
CONSTANT RATIO
STRUCTURAL CHANGE ONLY
TED/GWP RATIO
MAXIMUM CONSERVATION
0.5
0
1975
1985
2000
2020

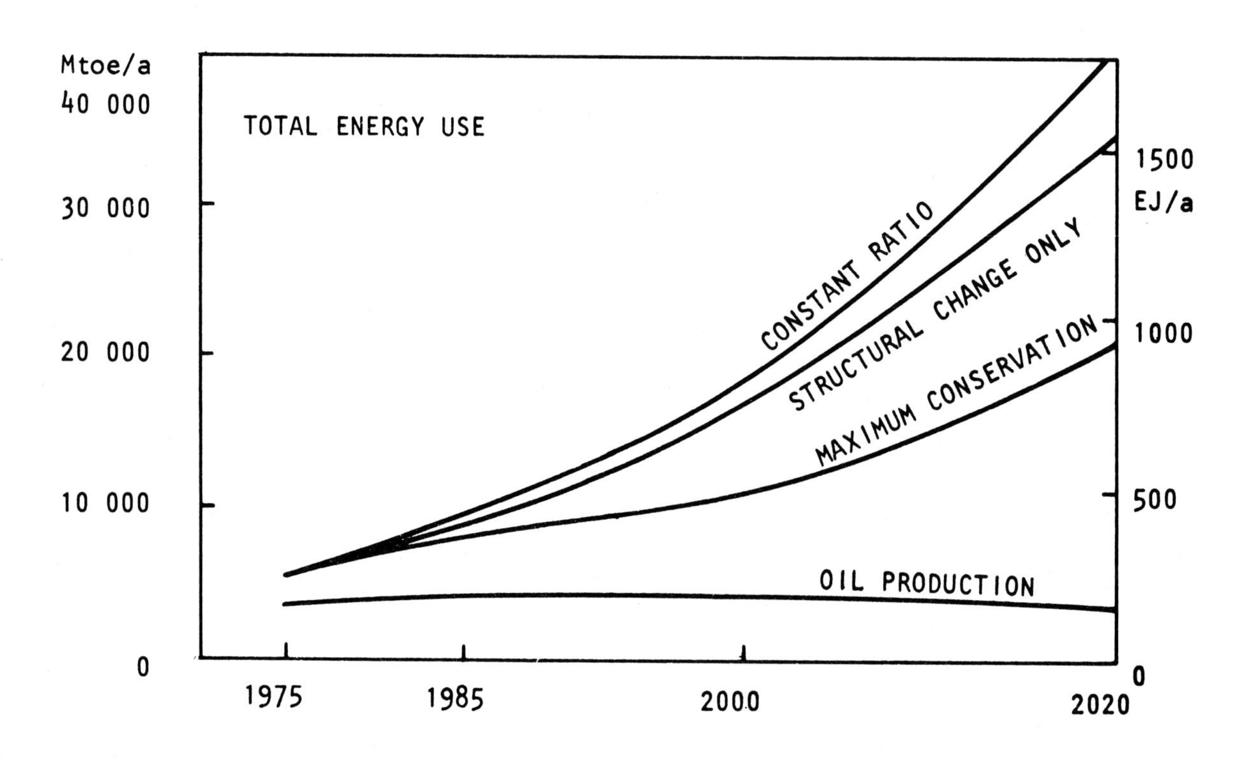

Mtoe/a
40 000
TOTAL ENERGY USE
30 000
20 000
10 000
0
CONSTANT RATIO
STRUCTURAL CHANGE ONLY
MAXIMUM CONSERVATION
OIL PRODUCTION
1500
EJ/a
1000
500
0
1975
1985
2000
2020

Three main conclusions can be drawn from the presented text:

1) The energy conservation potential measured by the TED/GWP ratio is 46 % in 2020. Part of the conservation, 17 %, is due to structural changes in energy use and the GWP, and the rest is conservation achieved through technological development and improved efficiencies.

2) The conservation potential increases with time so, that

9 %	is achieved	until	1985
33 %	"	"	2000
46 %	"	"	2020

This pattern of conservation timing coincides in a favourable way with the critical times when oil production cannot increase as before, notably in the 1980-1990:ies. At that time alternative energy resources are badly needed, and conservation offers such an alternative. After 2000 the demand again grows faster, the conservation potential being nearly fully utilized measured by todays knowledge. By then probably both new energy production and energy conservation methods will enter the scene.

3) The savings in the final consumption sectors 1 to 3 are mainly fuel savings. The fuel demand per GWP unit in these sectors is estimated to be 47 % of the 1975 value in 2020. The electricity demand per GWP unit will remain nearly constant, but the primary energy demand for electricity production per GWP unit will decrease by 33 % due to a higher efficiency of electricity production. The fuel demand for electricity production per produced kilowatthour is 37 % lower than 1975.

The result is sensitive to the future price ratio between nuclear electricity (later solar electricity) and fossile fuels. If nuclear electricity becomes lower priced in relation to fuels, the share of electrical energy will increase over the presented values, and the fuel demand will correspondingly decrease.

References

1) Energy Conservation in the International Energy Agency, 1976 Review. Organisation for Economic Cooperation and Development, Paris

2) World Energy Outlook. Organisation for Economic Cooperation and Development, Paris 1977

3) The National Energy Plan. Executive Office of the President. Energy policy and planning. April 29, 1977, Washington D.C., U.S. Government Printing Office Doc. nr. 040-000-00380-1. 103 pp.

4) "A Study of Improved Fuel Effectiveness in the Iron and Steel and Paper and Pulp Industries" by Gyftopoulos, Dunlay & Nydick. National Science Foundation, Washington D.C., U.S.A., March 1976

5) "Energy Conservation: Ways and Means", edited by Over and Sjoerdsma. Future Shape of Technology Foundation, publication nr. 19, 1974, the Hague, Netherlands.

6) "Efficient electricity use", edited by Craig B. Smith, Pergamon Press, 1976, sponsored by The Electric Power Research Institute, Cal., U.S.A.

7) "Potential Fuel Effectiveness in Industry" by Gyftopoulos, Lazaridis and Widmer. Ballinger Publishing Company, 1974.

8) "Encreased Energy Economy Efficiency in the ECE Region". E/ECE/883/Rev. 1. United Nations, Organisation for Economic Cooperation and Development, Geneve 1976.

9) "The Future of the World Economy" by W. Leontief. United Nations, New York, 1976.

Energy conservation in industry

1. Structure and trends of industry energy use

The complex structure of energy use in industry can be examined more readily by identifying the energy intensive industrial production processes. By combining global production statistics (U.N. statistics 1974) with estimated average specific energy demand coefficients the following picture can be drawn:

	World production	Specific energy demand		Energy demand	
	Mt/a	MJ/kg	kg oe/t[1)]	Mtoe/a	%
Iron and steel	705	32	800	560	25
Chemicals	-	-	-	300	14
Pulp and paper	150	40	1000	150	7
Cement	696	7	170	120	5
Aluminium	13	200	5000	65	3
Other industry	-	-	-	1005	46
Total	-	-	-	2200	100

The energy demand in the above table is given as total energy including purchased and internally generated electricity at a conversion factor of about 0,28 kg oe/kWh and including also internal or "captive" fuels like coke oven and blast furnace gas use in integrated steel mills, and spent liquor and hog fuel use in pulp and paper mills.

While each of the above industry branches must be treated according to its specific situation, many of the following characteristics can be attributed to the high-energy industries:
- capital intensive, thus long lead times for changes
- economy of scale important
- danger of institutionalization, often medium to low profitability

1) 1 MJ/kg = 24.8 kg oe/t

- cyclical market behaviour
- energy consciousness, energy being a major cost factor
- vivid national and international development exchange.

In the less energy intensive industries, included in "other", accurate information on energy has begun to accumulate only in recent years. Aggregate energy consumption trends for the whole industry are therefore not too reliable. The following table based upon reference 1.1 gives an approximation of specific energy demand trends and a forecast for the manufacturing industry in the United States.

Table 1 (based upon reference 1.1)
Rates of change of specific energy demand, percent per year

SIC		1954 to 1967	1967 to 1980
20	Food and kindred products	- 1.8	- 0.7
26	Paper and allied products	- 0.8	- 1.9
28	Chemicals and allied products	- 3.1	- 2.3
29	Petroleum and coal products	- 1.0	- 2.5
32	Stone, clay, and glass products	- 1.7	- 0.9
33	Primary metal industries	- 0.8	- 1.3
	Average of above	- 1.4	- 1.7
	All other manufacturing	- 1.0	- 1.1
	All manufacturing	- 1.6	- 2.0

Notes: Specific demand or energy/output ratio approximated as "energy consumed/value added in constant dollars". Petroleum and coal products are in our analysis excluded from industry and included in sector 5: "energy sector own use".

The study states that significant savings in energy use have been realized in the past. Energy use per unit of product declined at a 1.6 % average annual rate from 1954 to 1967, which was happening in a period of stable or declining energy prices. An extrapolation from 1975 to 2020 would give a reduction in energy use per unit of product of 50 %, according to this figure.

This source also gives the gross energy consumed by all manufacturing for 1947 and 1971 as 10 535 $\cdot$ 10^{12} and 19 864 $\cdot$ 10^{12} BTU:s respectively. During the same time the U.S. GNP increased from 468 $\cdot$ 10^{9} to 1108 $\cdot$ 10^{9} $ (constant 1972 dollars). The industry ED/GNP ratio thus decreased in 1971 to 80 % of the 1947 value corresponding to an annual percentage of - 0.9 (or 34 % in 45 years).

Although this extrapolation of past trends implies that the energy conservation goal for the industry sector estimated in chapter 3.2 (p.5) will almost be taken care of by technology development, factors like environmental demands, approaching limits of theoretical energy consumption, and depletion of raw materials suggest that this development needs to be pushed forward to some extent e.g. by energy pricing and research stimulating measures. With reference to profitability problems mentioned above, it is also considered essential to assist the energy intensive industry with capital for energy conservation investments. These industries tend to save their available financing capacity for strategic investments concerning new capacity, product quality, and pollution control. Government guaranteed loans (and even grants) are available in some countries and have been readily utilized under normal business conditions.

2. Examples of energy conservation potential in high energy industries

Generally the total conservation potential of 30-50 % is composed of

- 5-10 % immediate reduction as reaction to a substantial price rise like the oil crisis in 1973, achieved through improved energy management and minor modifications of existing processes
- 10-15 % in up to 5 years by process redesign and new equipment
- 10-20 % in 5 to 25 years by significant equipment and process changes involving heavy investments

according to reference (6).

In the iron and steel industry substantial differences exist between specific energy use in different countries (U.S. level in 1973 37 % over the level in Germany 1972), due to different age structure in the industry, and different energy price level. According to a U.S. study (ref. 4) specific consumption can be reduced by 11-17 % (12-17 Mtoe/a) from 1973 to 1983 by 1.5 - 9.5 billion US$ investment (the latter figure requiring government guaranteed loans).

In Finland the energy intensive industries in 1974 implemented or planned to implement before 1980 energy conservation investments amounting to $360 \cdot 10^6$ US$, yielding total energy savings of 1.24 Mtoe/a or 16 % of their total energy use (1.2). These investments became profitable as the fuel oil price increased threefold in 1974. Due to the economic recession, however, only part of this has been implemented as until now.

An estimate of medium term conservation possibilities in the Swedish pulp and paper industry (1.3) gave 0.58 Mtoe/a or 28 % saving (percent of external fuel) with $240 \cdot 10^6$ US$ investment, of which half was considered to be possible within 5 years.

In reference (7) the following percentages are given as long term specific fuel consumption conservation potential using technology existing in 1973, over 1968 practice, in U.S.A.:

Iron and steel	36 %
Petroleum refining	25 %
Paper and paperboard	39 %
Aluminium	32 %
Cement	43 %

For specific technical measures the reader is referred to the various references. A remark here is in place: improving energy economy in an industry plant by cogeneration of heat and electricity is in our report treated separately (sector 6 "electricity production" and Appendix 2), but is in most references treated as part of the industry energy economy.

References

1.1 Energy Consumption in Manufacturing. The Conference Board. Ballinger Publishing Company, Cambridge, Mass., 1974.

1.2 "Forest Industry Energy Study; Process Industry Energy Study." Summaries published by SITRA and Confederation of Finnish Industries and Central Association of Finnish Forest Industry. Helsinki 1976 (in Finnish).

1.3 "Energy supply of urban areas and heavy industry" SIND 1976:3. Liber Förlag, Stockholm 1976 (in Swedish).

Combined heat and power production in industry

1. Cogeneration potential

According to reference (6) 40 % of the total energy used in U.S. industry is in the form of process steam. If the same holds for the global industry energy demand of 1500 Mtoe in 1975, and a power to heat ratio of 0.2 for back pressure power generation is assumed, the total theoretical cogeneration potential would be 2200 TWh of electricity or 82 % of industry electricity demand (Table 1, p. 18). This "theoretical" potential can of course be considerably larger if technologies with higher power to heat ratio is considered, e.g. gas turbine or diesel cogeneration systems. Obviously, however, a large part of this potential cannot be utilized economically for various reasons, e.g. the following:
- too small units
- high back pressure
- fluctuating heat load

Today the most frequent obstacle is low electricity price. If electricity is priced according to condensing power generation cost from the same fuel as cogeneration power, the back pressure electricity will be competitive at 5 to 10 MW_e plant size (Ref. 2.1), provided that the same depreciation factors are applied for both cases.

2. Present utilization of potential

Traditionally inplant generation of back pressure power is best developed in the pulp and paper industry, especially where power prices have been high (Eastern, Central and Southern U.S.A., Federal Republic of Germany, Austria, Finland). In these areas back pressure power has supplied around 50 % of the total electricity demand of this industry.

According to various sources the amount of industrial back pressure generated electricity for selected countries was as follows in 1973:

	TWh	Percent of total industry electricity demand
U.S.A.	100	12
Fed.Rep.of Germany	24	15
Finland	5.6	31
Sweden	3.9	10

A guess on the world total as 200 TWh/a gives a 9 % utilization of the above potential.

3. Cogeneration in chemicals and petroleum refining

In a study for the U.S. Federal Energy Administration (ref. 2.2) the fuel saving potential in the U.S.A. has been evaluated for the chemical, petroleum refining, and pulp and paper industries, by applying three technologies for cogeneration, namely steam turbine, gas turbine, and diesel topping technology. The fuel savings[1)] that were thermodynamically possible in comparison to present operations were estimated to 14, 36, and 53 % respectively for the three technologies. The economic potentials for implementation in 1985 amounted to 30-50 % of the thermodynamic potential, and could be increased to 70 % with federally guaranteed loans for half the investment. A still better economic potential was obtained if the ownership of the inplant generation facilities would be by electrical utility companies.

When discussing the merits of different technologies the report states that gas turbine and diesel topping requires gaseous or liquid fuel and represent less proven applications, but the fuel saving potential is superior due to the favorable power to heat ratio, although the efficiency is lower than steam turbine topping. (In the future gas turbines may be fed by pressurized fluidized bed combustion chambers, whereby solid fuels can be utilized. See figure 1).

---

1) Savings in % is compared to total energy requirements of the industries in question. Fuel saved is calculated as difference between fuel demand of utility generated power and inplant generated power.

The inplant generation would in the cases of gas turbine or diesel topping not only render the investigated industries self-sufficient in electrical power, but would also allow power export; in the most favorable case (maximum economic diesel potential with federal economic incentives) 1200 TWh/year in 1985 equivalent to 35 % of U.S. utility generation. Assuming a load factor of 7000 h/year this energy corresponds to a electricity generation capacity of 170 GW.

The study discusses also technical, operational, and nontechnical barriers to implementation of inplant generation at such a large scale.

As a general conclusion the inplant generation is considered as an extremely important strategy for energy conservation in the U.S.A.

References

2.1 "Back pressure power generation in the Nordic countries". Report for Nordic Industry Foundation. The Finnish Energy Economy Association, Helsinki 1977 (in Swedish).

2.2 A study of inplant electric power generation in the chemical, petroleum refining, and paper and pulp industries. FEA/D-76/321, June 1976.
NTIS, U.S. Department of Commerce. Virginia.

WORLD ENERGY DEMAND TO 2020

I. J. Bloodworth, E. Bossanyi, D. S. Bowers, E. A. C. Crouch, R. J. Eden, C. W. Hope, W. S. Humphrey, J. V. Mitchell, D. J. Pullin and J. A. Stanislaw
Energy Research Group, Cavendish Laboratory, University of Cambridge

Contents

1 **The outlook**
Summary
Questions for discussion

2 **Assumptions and methodology**

3 **World energy demand**
3.1 Primary energy demand
3.2 Oil demand
3.3 Fuel substitution
3.4 Conservation, efficiency and price response

4 **Regional energy balances**
4.1 OECD
North America
West Europe
Japan, Australia, New Zealand
4.2 Centrally planned economies
USSR and East Europe
China and centrally planned Asia
4.3 Developing group
OPEC developing
Non-OPEC developing

Units and conventions

Most exhibits display energy both in Standard International Units, using the GigaJoule (GJ) or the ExaJoule (EJ), and in Tonnes of Oil Equivalent (TOE), using the higher calorific value for oil

1 ExaJoule (EJ) = 10^{18} J = 10^{9} GJ
1 ExaJoule = 22·7 million TOE
1 TOE = 44 GJ

Nuclear and Hydro primary energy are defined to be equal to the corresponding electrical energy generated divided by (0·35).

1 ExaJoule of Primary Nuclear at 70 per cent load factor corresponds to approximately 16 GW(e) of generating capacity.

Net Electricity Output is defined to be equal to (0·85) times Gross Output and is equal to (0·85) (0·35) times Primary Energy Input.

Exhibits

1 World primary energy demand projections
2 Summary of projections to 2020
3 World energy demand and supply
4 World primary energy demand by fuel
5 WEC world regions
6 Energy demand methodology
7 World energy balance table 1972
8 World and world groups total GNP
9 World population and GNP shares
10 Guidelines assumed for potential energy supply

11 World energy demand: unconstrained projections
12 World energy demand: percentage shares
13 Price response of world energy demand
14 World energy demand and potential supply
15 World transport energy: percentage shares
16 World oil demand
17 World oil demand by sectors
18 Shares of world primary energy demand
19 World energy demand and potential supply

20 OECD primary energy demand projections
21 North America: energy demand
22 West Europe: energy demand
23 Japan, Australia, New Zealand: energy demand
24 Centrally planned group: primary energy demand projections
25 USSR and East Europe: energy demand
26 China and C.P. Asia: energy demand
27 Developing group: primary energy demand projections
28 Energy demand in the developing regions
29 OPEC developing group: energy demand
30 Non-OPEC developing group: energy demand

Note by the Conservation Commission

As indicated in the Foreword and in a footnote in this Report, electricity may be converted to heat and work at higher average efficiency than other forms of secondary energy. In certain current economic studies a Joule of electrical energy is regarded as having 2 to 3 times the average value of a Joule of energy in fossil fuels for conversion to heat and work. On this basis electricity's share of the total value of secondary energy is about the same as the share of total primary energy used in its production.

FOREWORD

This Executive Summary is a revised version of the report on energy demand prepared by the Energy Research Group for the Conservation Commission and presented at a Round Table discussion at the Tenth World Energy Conference in Istanbul. The authors wish to acknowledge the valuable advice received from members of the Conservation Commission and from many of the delegates at the World Energy Conference.

The revisions indicated below have been carried out in detail in the Full Report to be published in the spring of 1978. In this revision of the Executive Summary the main points are noted in context, but the more detailed alterations and additional explanations are not included. The main revisions that have been made in response to the discussion at the World Energy Conference in Istanbul and comments received by the authors are the following:

(i) The three basic scenarios H4 (high growth), H5 (high growth high conservation), and L4 (low growth) are given more equal emphasis to take account of the uncertainties in the scenarios and to avoid the impression that any one of these represents a forecast.

(ii) Scenarios (fast development) in which the Developing Regions have higher economic growth than in the high growth scenarios are described in more detail.

(iii) A number of variants on the basic scenarios are introduced in addition to the fast development scenarios. One is a transition scenario in which high growth to 2000 is followed by very low growth from 2000 to 2020. Others involve faster interfuel substitution than that used in the basic scenarios.

(iv) The upper limits of 50 or 60 per cent nuclear electricity as a fraction of total electricity are removed.

(v) The demands for primary energy in each region are indicated by ranges of values to emphasise the uncertainties in such projections.

(vi) A number of details of the projections have been corrected or revised. These include some minor changes in economic growth rates, an increase in the projected demand for gas in OPEC, and higher growth for electricity demand in West Europe.

Richard J. Eden 15th December 1977

Energy Research Group, Cavendish Laboratory, Cambridge, England.

1 The Outlook

Summary

Energy demand and conservation
World energy demand in the year 2020 is expected to be between three and four times present consumption if average economic growth is similar to that achieved in the past forty to fifty years and there are vigorous and successful measures to improve the efficiencies with which energy is used.

Primary energy
Even with vigorous energy conservation there will be an increased demand for all forms of energy, but an initial consumer preference for oil may delay the widespread substitution of other fuels so long as oil supplies are available.

Oil
World demand for oil will increase until the period 1985 to 1995 when consumption will become constrained within the limits set by potential oil supply. Conventional sources of oil should be adequate for premium uses such as transport and chemical feedstock beyond the end of the century, but this extended period for the premium uses of oil implies a decline in the use of oil for heating beginning in the 1980s, and an increasing use of electricity for transport is expected after the year 2000.

Substitution
The decline in heating oil may be achieved partly through improved efficiencies but it will also require widespread substitution from other energy sources. The magnitude of this substitution is an important result of this study.

Coal
Much of the new demand for fossil fuel will be met directly or indirectly by coal, and world coal demand is likely to increase rapidly after the mid 1980s. Coal will be required in industry both for direct use and for making synthetic fuels, and to balance the nuclear component in electricity generation. World coal demand in 2020 is expected to be between 4 and 5 times its present level.

Gas
Some of the future demand for fossil fuel can be met by increased supplies of natural gas, but supplies from some existing sources will begin to decline so this would require the transport of gas from new sources to major markets. In some markets imported gas or synthetic gas will be in competition with electricity as a replacement fuel for oil and to meet increasing energy demand.

Electricity
The share of the demand for secondary energy* that is met by electricity is projected to increase from its present value of 10 per cent to nearly 20 per cent in the year 2020. This would require the primary energy for the generation of electricity to increase more than sixfold so that its share of the total world primary energy would rise from 25 per cent to nearly 40 per cent.

*the share of useful energy would of course be higher than that for secondary energy

Nuclear

The projected increase in world electricity demand could not be met without a major contribution from nuclear power. The nuclear share of electricity output could be almost 45 per cent by the year 2000 and 65 per cent by 2020. The world total for nuclear generating capacity implied by these percentages would lie between 1300 and 1900 GW(e) in the year 2000 and between 3200 and 5500 GW(e) in the year 2020, depending on growth and the load factor achieved. The balance of electricity generation would need to come mainly from coal and hydropower.

Renewable energy sources

The current 15 per cent share of world primary energy met by renewable sources (hydro, wood and solar) remains nearly constant in our projections. This still implies a substantial increase in absolute terms, and hydropower is assumed to increase fourfold, whilst solar energy increases from almost zero until, in 2020, it provides as much useful energy as that derived from electricity at the present time.

Global distribution

The historical trend shows that the Developing Regions have increased their shares of world energy consumption and this trend is continued in our projections. The Developing Regions now include 50 per cent of the world's population but use only 15 per cent of the annual energy. By the year 2020 their share of population is expected to be about 60 per cent, and their share of energy could be about 25 per cent if there is high economic growth.

Fast Development scenarios for Developing Regions

Fast development scenarios have been examined in which the Developing Regions are assumed to have economic growth at an average annual rate of 6.3 per cent from 1975 to 2020, as compared with 5.3 per cent in the high growth scenario. This fast growth assumption has been combined with alternative assumptions of high and medium conservation and studied with various alternative scenarios for OECD. This increases the Developing Regions' total energy demand by about 30 per cent in year 2000 and by about 70 per cent in 2020.

International trade

Much of the additional demand for gas and coal will need to be met through international trade. By 2020 the international trade in these two fuels could be as large in volume as the present international trade in oil. The trade in oil is expected to increase during the next ten years but is likely to begin a slow decline near the end of the century. Our projections do not indicate substantial energy trade between the Centrally Planned Regions and the rest of the world.

Energy conversion and energy prices

With increasing requirements for conversion to electricity and to synthetic gas and synthetic oil, it is expected that the percentage of primary energy consumed by the energy industries during conversion will rise from its present value of 24 per cent to about 35 per cent by 2020. The increased amount of conversion will lead to increased average costs of energy supplied to the final consumer, additional to those arising from any increased costs of primary fuels. The additional conversion costs will be partly offset by an increase in the average amount of useful energy that is contained per unit of energy supplied to the final consumer. No forecast is made about future real energy prices.

Energy and economic growth

The energy demand projections in this report are related to alternative assumptions of 'low' (3·0 per cent per annum) and 'high' (4.1 per cent) future world economic growth. With low growth the average per capita income would nearly double by 2020 and with high growth it would nearly treble. If energy and income were to maintain the same relation as over the past fifty years, the resulting projections would lead to an energy demand in the year 2020 ranging from more than four times to more than six times the present consumption.

'Low' economic growth is associated with normal conservation responses that imply substitution of other goods and services for energy as energy prices rise relative to the general price level. This substitution is most marked in the projection of energy for transport, which is significantly below the historical trend, partly due to greater efficiency arising from the expected scarcity of oil with respect to other fuels, and partly due to approaching saturation of demand for private motoring in major consuming countries. The resulting low growth 'oil-constrained' projection (L4) for world primary energy demand in the year 2020 is 840 ExaJoules, or about three times the present consumption. This level of demand lies within the range of potential world energy supply estimated for that year.

'High' economic growth is associated with similar assumptions to those used for the low growth projection, but it includes also the essential additional assumption of 'energy minimising' conservation that leads to significant long run improvements in the efficiency of industrial and domestic use of energy that go beyond the normally expected consumer response to higher energy prices. In effect, this implies that other factors are substituted for energy to yield equivalent output or satisfaction, and these lead to a changed or different development pattern in which economic growth is associated with correspondingly reduced energy growth. One of the main conclusions of this study is the necessity for such 'efficiency improvements', and an indication of their magnitude and timing that permits the resulting high economic growth 'energy constrained' projection (H5) to be associated with an energy demand of 990 EJ in the year 2020, which is within the estimated range of potential world energy supply.

Regional energy balances

The efficiency improvements implied by the high growth energy constrained projection (H5) are necessary in all major world regions if an energy balance is to be achieved by the projections to 2020. This conclusion applies not only to OECD which currently dominates world energy demand but also to the Developing Regions and the Centrally Planned Economies. In order to achieve an energy balance in 2020 it is also necessary to assume a vigorous development of indigenous energy resources in each region and a willingness to export from those regions that are well endowed with energy resources.

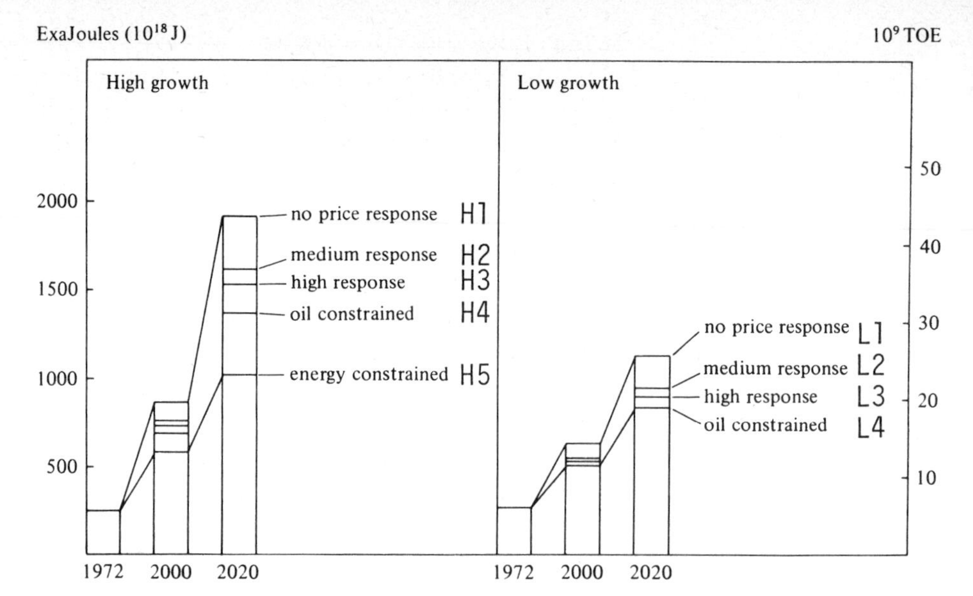

Exhibit 1. World primary energy demand projections

Potential world primary energy demand EJ		
Year	1975	2020
World total consumption in 1975 including 26 EJ wood fuel	276	ExaJoules
Unmodified historical trends (at constant growth rates)		
1960–75 trend continued		1952
1933–75 trend continued		1770
1925–75 trend continued		1242
Projections found from GNP assumptions		
Low economic growth (3·0 per cent p.a.)		
L1. No price response		1128
L2. Medium price response		953
L3. High price response		897
L4. High price response plus oil constraints		840
High economic growth (4·2 per cent p.a.)		
H1. No price response		1831
H2. Medium price response		1546
H3. High price response		1455
H4. High price response plus oil constraints		1322
H5. High price response plus energy constraints		990
Potential world energy supply in 2020 (see exhibit 10)		820-1010

Exhibit 2. Summary of projections to 2020

Units: ExaJoules = 10^{18} J = 0·023 Gigatonne oil equivalent (see notes on units and conventions on page ii)

Questions for discussion

A number of questions were raised by the authors for discussion by the WEC in Istanbul, and others were raised by those attending the Conference. Few of these questions can be answered with certainty and on some there is a wide diversity of views. Section 2.2 of the Full Report includes the detailed comments on these questions. It is not possible to represent all viewpoints in this Executive Summary, but the following notes are designed to illustrate some of the central issues and areas of uncertainty.

1 Limits to growth

With low economic growth we expect world energy demand by 2020 to be three times current consumption. High economic growth could be associated with a fourfold increase in energy demand by that year if there were unprecedented long term improvements in the efficiency with which energy is used. The necessity for these efficiency improvements arises from our assumption that energy supply in the future will be more inelastic than in the past. Is this assumption correct? Is it likely that difficulties in increasing energy supplies will limit the scope for growth of energy demand?

2 Energy efficiencies

If future energy demand is limited by potential supply, can high world economic growth be maintained? Our results suggest that continued high economic growth would be feasible if there were cumulative long run substitutions of other goods and services for energy, so that energy growth slows down compared with economic growth. Could the required improvements in efficiency be a natural response to higher prices and an expectation of energy scarcity? Or would the required level of response develop only under crisis conditions? Would government intervention and international agreements be necessary?

3 Energy prices and the response of consumers

Our projections have assumed that the average real price of energy to the final consumer doubles over the next twenty-five years. Can such a change evolve smoothly and naturally through normal market forces as a result of expectations of scarcity? Alternatively, if higher prices are an essential component of energy conservation, will they need to be administered, either by producing governments through world trade in energy, or by consuming governments through taxation?

4 Long term improvements in energy efficiencies

Are the improved efficiencies required in the use of energy for transport, industry and the domestic sector so urgent that they will require government intervention through research, regulations, and standards. Or can one rely on technological progress and future energy prices to achieve the same response?

5 Time for change

In directing our attention to the year 2020 our projections lead to relatively low growth in energy demand compared to economic growth over the next decade or two. This is because we have assumed that improvements in energy efficiencies required in the long term will begin to be adopted now. Is this likely to happen? Are our assumptions about long lead times for energy conservation unduly pessimistic? Or are our assumptions about the amount that could be achieved too optimistic?

6 Electricity, coal and nuclear power

The projections give a world demand for electricity in 2020, which is between six and seven times its present value, and may be compared with a corresponding increase of nearly four in total energy demand. This relatively faster growth in electricity demand raises a number of questions: Could we realistically have assumed a different growth rate for electricity? Could the growth of nuclear be realistically varied, for example, by assuming fast growth for nuclear process heat? Could the growth in coal supply be greater than that described in the coal study?

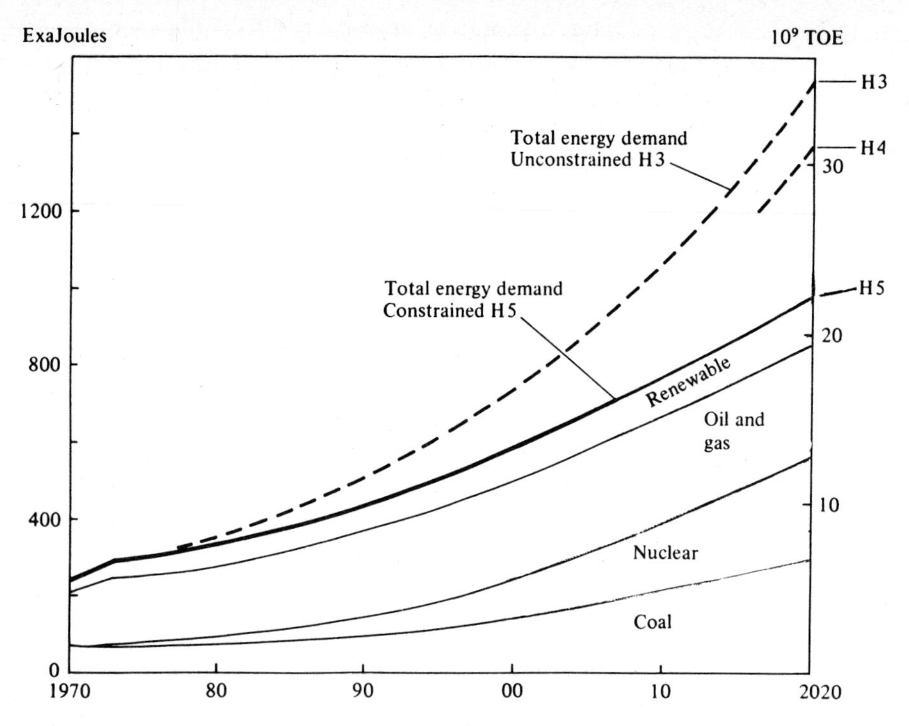

Exhibit 3. World energy demand and supply
High economic growth

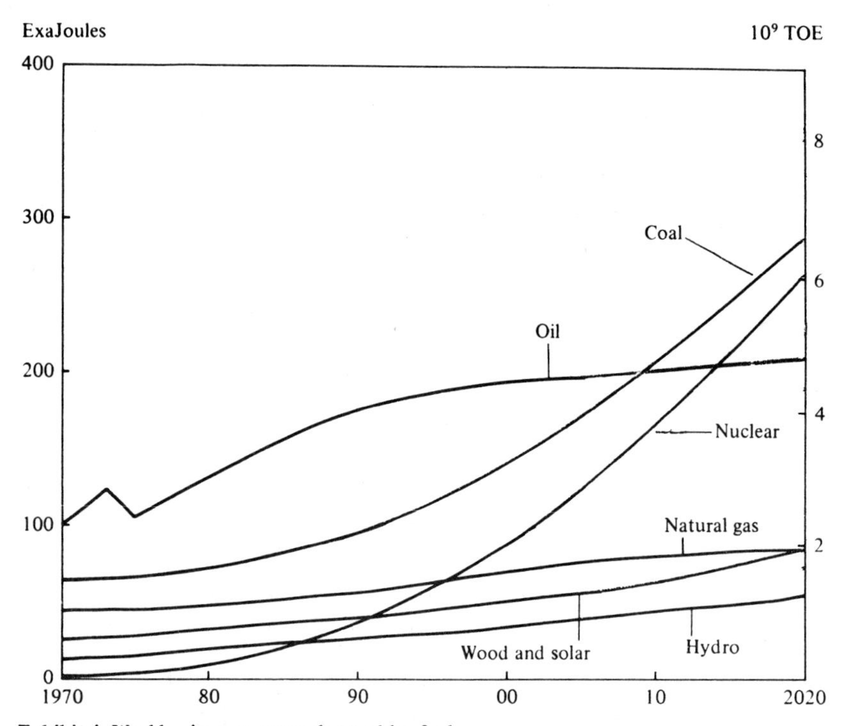

Exhibit 4. World primary energy demand by fuel
Constrained high growth H5

7 Limits on nuclear

The assumption of the draft Executive Summary that electricity generated by nuclear power remains less than 60 per cent of total electricity produced has been relaxed in this revised Executive Summary. The world-wide average now rises to about 65 per cent by 2020 compared with 57 per cent in the draft report. The main reason for the change is that the pressure on world coal production indicated by the projections would raise coal prices and affect the economic balance in electricity production. Even though the resulting demand for nuclear power in the projections remains below the figures used in the nuclear resources study, it still raises the question whether such an increase could be feasible.

8 Renewable energy sources

We have assumed that hydro-electric power could increase fourfold by 2020. No supply limits have been imposed on wood fuel or solar energy, and their supply is driven entirely by demand. In our projections the demand for wood fuel remains nearly constant, but solar energy increases so that it provides as much useful energy in 2020 as electricity does at present. However, solar energy remains a small part of energy supply and in 2020 it is still less than 5 per cent of the world total. Have we underestimated the potential demand for solar energy and the rate of market penetration that is likely?

9 International trade

The projections imply a greatly increased international trade in liquefied natural gas and in coal. Will the physical facilities be developed in time and will investors have confidence in their access to international markets, both as exporters and importers, to the extent necessary to bring in the necessary development of supplies?

10 Interfuel substitution and energy conversion

The projections show substantial substitution of coal for oil in the supply of heat. Will this be costly and will it have environmental and social consequences that need to be taken care of in advance? Would consumers choose to use more electricity, synthetic gas or synthetic oil to avoid the problems associated with the direct use of coal and would this choice be limited by the large capital investment required?

11 Alternative scenarios

We have examined three basic scenarios and a number of important variants to indicate the magnitude, variety and timing of energy problems that must be resolved if reasonable world economic growth is to be maintained. The basic scenarios include H4 with high growth and moderate conservation, H5 high growth and high conservation and L4 low growth and moderate conservation. The variants include scenarios in which the Developing Regions have higher economic growth than that assumed for the high growth scenarios H4 and H5. The variants also include a transition scenario in which an initial period of high economic growth is followed by a period of very low growth.

12 Our conclusion

High world economic growth during the period to 2020 would require early and successful action towards energy conservation and restraint accompanied by vigorous development of all forms of energy supply. These requirements for increased conservation and supply are illustrated in Exhibits 3 and 4. If less energy conservation or less growth in supply was achieved, high economic growth would still be feasible for an initial period, perhaps for ten years and possibly for twenty, but at the end of this initial period, severe shortages of energy would develop rapidly. The prospect of severe energy shortages would be anticipated by increases in the price of energy before they developed, possibly within the next ten years. We therefore ask whether a new surge in energy prices would so disrupt the economic system as to lead to a subsequent prolonged period of low economic growth or instability? If the answer is no, it can only be because it is believed that under crisis conditions major adjustments on conservation and supply would take place rapidly. But if they can take place rapidly under crisis conditions, it seems reasonable to suppose that they could take place less rapidly through adequate planning under present conditions. The essential question is whether the speed of change in energy use and the growth in supply will be sufficient for high economic growth to be maintained. In this report we have indicated the magnitude and timing of the changes that would be required.

In this revised Executive Summary for illustrative purposes many of the exhibits are based on the high growth, high conservation scenario H5. This projection H5 does not represent a forecast since it would require successful achievement of an unprecedented level of energy conservation. If less conservation was achieved the projected demand would lie between H4 and H5. However, if the necessary additional energy production was not achieved the associated economic growth would be correspondingly reduced. The detailed figures for fuel demand in individual world regions are uncertain and some of this uncertainty is indicated in this revised Executive Summary in those exhibits that have been revised to display the ranges of values for energy demand that would be associated with the uncertainties in the scenarios. The details of the three basic scenarios H4, H5 and L4 and the variants are not presented in this revised Executive Summary, but will be given in the Full Report (to be published in 1978).

Nevertheless, many of the conclusions remain valid for the whole range of scenarios considered, though the magnitude and the timing of periods of energy scarcity and problems of substitution may change. The changes in presentation in this report and in the Full Report do not lead to any less emphasis on the long term need for energy conservation, but they do give greater emphasis on the importance of the vigorous development of all forms of supply. The main conclusion remains firm: high world economic growth during the period to 2020 would require early and successful action towards energy conservation and constraint accompanied by vigorous development of all forms of supply.

2 Assumptions and methodology

The problem

The main question at issue in this report is the possible range of response in energy demand to an impending scarcity of oil relative to other energy carriers, and the possibility that the average real price of energy to the consumer may rise substantially relative to other goods and services. The method adopted aims to describe the essential features of the problem through broad, simple assumptions, which are consistent with each other, and then with the help of computer models to identify those assumptions that are critical to the main question and to analyse the consequences of varying these critical assumptions.

The 'projections' are only *conditional* – conditional on the assumptions intended to contribute to the discussion of a limited problem. The projections of total energy demand and the distribution of demand between different fuels or energy carriers are uncertain, but they are believed to provide typical 'scenarios' for high or low economic growth that illustrate the magnitude of the problems of energy supply and energy conservation that must be faced during the coming decades. The projections are based on the world regions listed in Exhibit 5, and the methodology is structured in Exhibit 6.

Group	Region
OECD (1, 2, 3)	1. North America 2. West Europe 3. Japan, Australia, New Zealand
CP (4, 5) (Centrally planned)	4. USSR/East Europe 5. China and centrally planned Asia
Developing (6 to 11) OPEC developing (6) Non OPEC developing (7 to 11)	6. OPEC 7. Latin America 8. Middle East and North Africa 9. Africa South of the Sahara 10. East Asia 11. South Asia

Exhibit 5. WEC world regions

Constant assumptions common to all cases

1 Basic data on energy demand in each world region by fuel (including wood fuel) and sector in 1972, which is taken as the base year for the projections. This is illustrated and summarised in Exhibit 7.

2 Regional variations in GNP (gross national product) corresponding to low (3·0 per cent per annum) and high (4.1 per cent) world economic growth for the period 1975–2020. These are summarised in Exhibit 8 and their effect on the distribution of world total GNP is shown in Exhibit 9. In the low growth case the world average GNP per capita in 2020 is nearly twice today's level, and in the high growth case it is nearly three times (we assume the UN median projection for population to 2000 and extend it to 2020 at the average growth rate for 1990–2000).

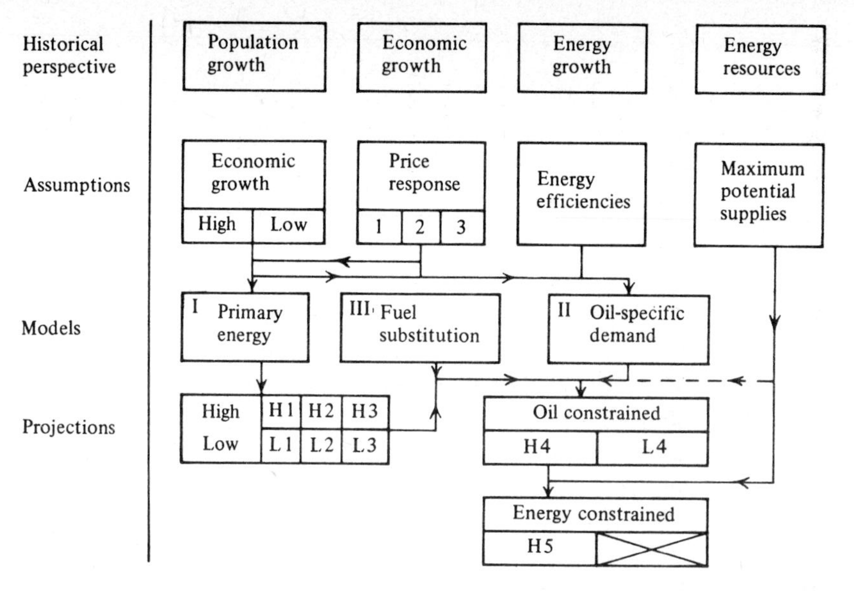

Exhibit 6. Energy demand methodology

3 No political, economic or physical disasters. Any of these might alter the smooth extrapolation to the year 2020, though it may be noted that the average economic growth for the past 50-year period containing the world depression of 1929–39 is higher than that assumed in the low growth projections.

4 A preference for the use of indigenous energy supplies (if they are available) in the world regions shown in Exhibit 5, representing the idea of a cost penalty for transporting fuels.

Constant assumptions relating to fuel substitution

5 A preference against the manufacture of synthetic fuels such as oil and gas until the total world demand for those fuels enters the range of values where a potential scarcity of supplies is expected.

6 In regions 1 to 3 nuclear power is assumed to grow rapidly from 1972, in region 4 from 1977, and in regions 5 to 11 from 1985.

	Coal	Oil	Gas	Elec.	Heat†	Wood	Sol.	Nucl.	Hydr.	Total
Transport	1·9	41	0	0·5	0	–	–	–	–	43·4
Industry	22·3	21·7	18·9	9	3·5	5·2	–	–	–	80·6
Domestic	8·1	17·9	12·7	7·6	1·3	20·9	–	–	–	68·9
Feedstocks	0·1	9·7	0·7	–	–	–	–	–	–	10·5
Total final consumption	32·4	90·3	32·3	17·1	4·8	26·1	–	–	–	203
Electricity generation	27·4	15·1	9·1	−20·3	−5·8	–	–	1·6	13·8	40·9
Synthetic gas	2·7	1·9	−3·8	–	–	–	–	–	–	0·8
Synthetic oil	0	−0	–	–	–	–	–	–	–	0
Energy sector own use	3·1	6·8	8·4	3·2	1	–	–	–	–	22·6
Primary energy input	65·6	114·1	46	–	–	26·1	–	1·6	13·8	267·2
Indigenous supply	65·8	115·2	46	–	–	26·1	–	1·6	13·8	268·5
Net imports*	−0·2	−1·1	0	–	–	–	–	–	–	−1·3

* And items unaccounted.
† 'Heat' cogenerated with electricity.

Exhibit 7. World energy balance table 1972
Units: ExaJoules = 10^{18} J

7 Rules for substitution between fuels as follows:

(*a*) A preference order within each economic sector for particular fuels (see Exhibit 7). This represents the differences in convenience and cost to the user between one fuel (or electricity) and another, and takes account of the world supply potential for different fuels in future years.

(*b*) A concept of inertia, restricting the rate at which one fuel may be substituted for another within a sector, indicating that plant for using a new fuel will penetrate a market only slowly as methods change and old plant is retired. The modelling assumptions restrict both the rate of growth and the rate of decline of each fuel for the final demand sectors. A faster rate of growth is permitted for fuels in the energy supply sectors, but here also the rate is related to the potential development of new plant.

Assumptions 5 to 7 give an intuitively plausible set of rules for projecting interfuel substitution in the absence of detailed information about costs and cross elasticities for fuels in the future. A degree of judgement is involved in deciding plausible preference orders for individual fuels in different sectors, but these are either explicitly discussed (as with oil for transport, or nuclear for electricity), or their effects cancel out, and do not strongly influence aggregate demand for individual fuels. The rules take account of the substitution of commercial energy for wood fuel and farm waste, which is of particular significance in developing regions.

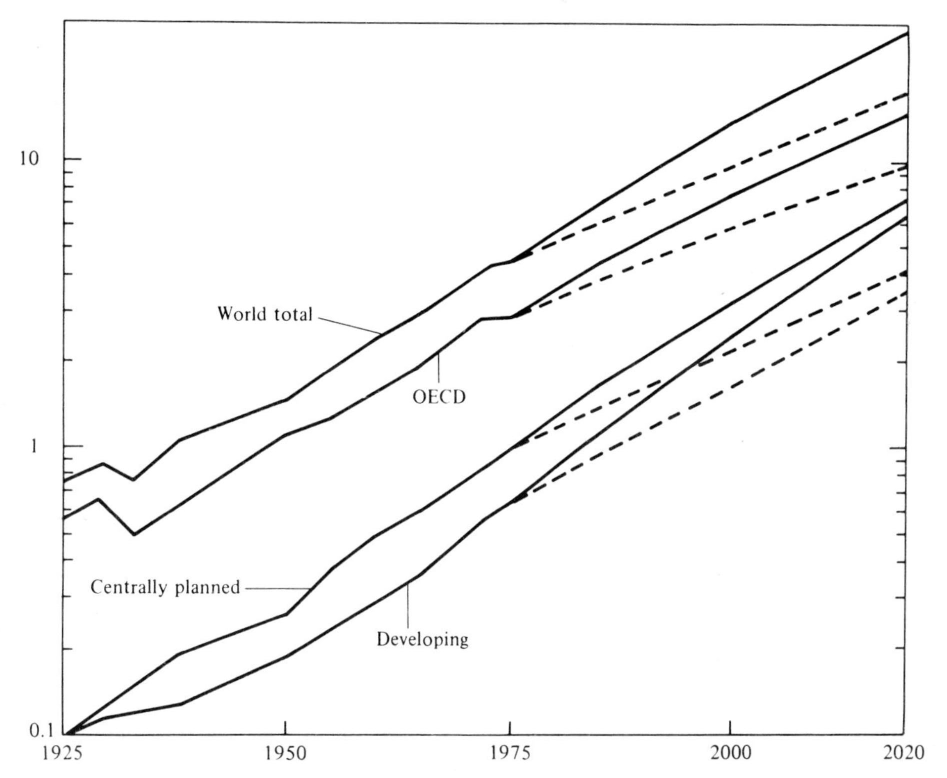

Exhibit 8. World and world groups total GNP/10^{12} U.S. (1972)
High growth, ———; low growth, – – – – –

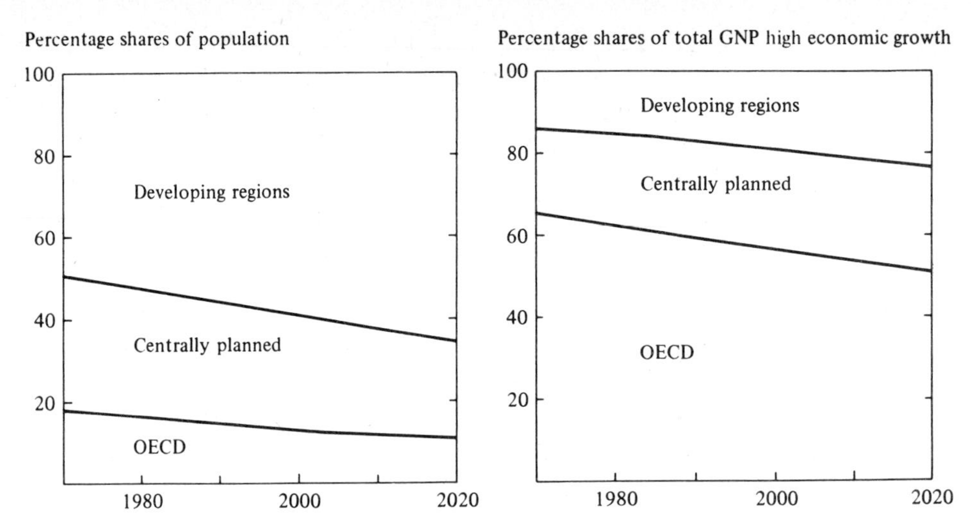

Exhibit 9. World population and GNP shares

Critical assumptions

The critical assumptions concern the substitution of ' non-energy ' for energy, within the category of goods and services that are counted in GDP (the subtler problems of qualitative satisfaction of consumers are not considered here though they may be implicit in some of the assumptions). The significance of the assumptions is illustrated below by brief reference to their consequences:

No price response

8 If there was no concern about future energy supplies or costs we could, on the basis of historical relationships, assume that world energy demand would grow a little faster than GDP (slightly higher in developing countries, slightly lower in fully industrialised countries). The consequent factor of increase from 1975 (1·00) in world energy demand would be:

Projection	Average economic growth 1975–2020	Primary energy ratio 1975	2000	2020
H1	4·2 per cent per annum	1·00	3·18	7·00
L1	3·0 per cent per annum	1·00	2·31	4·12

Under this assumption, only an average rate of economic growth to 2020 of less than 3 per cent would fall within the range of expected supplies. Such a growth would be consistent with a growth of 5 per cent for 10 to 15 years followed by a prolonged depression in which economic growth was no faster than population growth. This would be the reverse in time of the growth pattern over the past 50 years.

Constant ' normal ' price response

9 Assumption 8 might be varied by allowing for energy conservation additional to that already implicit in historical trends. One way in which this conservation might be estimated is by assuming a price increase over and above that which has already occurred by 1977, together with an elasticity of response by the final user of energy. To test the effects of this sort of change, we have considered combinations of price increases and elasticities which together yield reductions in energy use by the year 2020 of between 15 per cent and 21 per cent below the cases L1 and H1. For illustration we have assumed price increases that raise the average real price of energy for the final consumer to 1·7 or 2·2 times the 1972 energy price, together with an ' elasticity of response ' of minus (0·3). These price increases give projections L2, H2, and L3, H3, which (by 2020) are respectively 15 per cent and 21 per cent below the correspond-

ing energy demand in L1 and H1. Under these assumptions, world energy demand associated with 'low' (3·0 per cent) economic growth, would rise from 268 EJ in 1972 to a value in 2020 in the range 830 to 970 EJ, which is within the estimated range of supply. No significantly higher economic growth would be feasible unless greater supplies were available at these prices, or unless the change in energy demand exceeded these 'normal' responses.

Premium uses for oil

10 Assumptions 8 and 9 follow from relatively simple ideas about energy as a whole. However, we cannot sensibly avoid considering individual fuels, particularly oil. If the demand for oil increased at the same rate as energy, even in the low growth case L3 oil demand would increase over three times by 2020 and would be likely to exceed the maximum potential supply before the year 2000. We therefore assume that scarcity of oil will increase its price relative to other fuels. The cost of using any fuel other than oil in many forms of transport is likely to exceed the cost of similar substitution for oil in domestic heating or industrial heating and steam raising, so we assume that oil would be gradually priced out of the heat market (except for specialist applications), in line with our 'substitution' assumptions (7 above). The use of oil as a chemical feedstock is assumed also to maintain its premium value, as with transport.

11 We assume that the demand for energy in transport in the long run is modified in response to the relative price increase for oil by two developments:

(*a*) The 'saturation' of the demand for private motoring in the more wealthy countries will have a lower saturation level than might otherwise have been expected. At the margin, motorists will substitute other forms of transport and other forms of consumption for private motoring.

(*b*) Technical improvements, which are already targeted (for example, in the U.S.A.), and changes in load factors, will in the long run significantly increase the efficiency with which oil inputs deliver ton miles or passenger miles of transport.

The effect of these assumptions has been studied in sub-models of the transport sector of the main regions. While much of the detail must be speculative, the overall effects seem clear. Compared to the total energy for transport, the share required for road transport will fall (Exhibit 15), though there is likely to be a substantial increase in the transport energy requirements of the developing regions. The total demand for oil will start to become 'constrained' (Exhibits 16 and 17), within the period 1985 to 1995.

General long run responses

The special case of oil as a relatively scarce fuel has led us naturally to expect its allocation to premium uses such as transport and chemical feedstock. In the transport sector we have identified a variety of long run adjustments to the increasing relative cost of oil that could enhance already approaching saturation of demand and accelerate the development of better technical efficiencies, better operational efficiencies (such as load factors) and some modal changes. All of these contribute to changed relationships between oil demand, transport activity, and economic growth.

We cannot be so specific about the potential for similar long run adjustments in the domestic and industrial sectors, but we can reasonably assume that such adjustments could take place in the face of an increasing scarcity of energy itself and an increase in the price of energy compared to all other goods and services.

With all other assumptions unchanged, and keeping within the estimated range of supply, an energy demand of the order of 990 EJ by 2020 (compared with 276 EJ in 1975) could be associated with high economic growth averaging 4·2 per cent per annum (instead of low at 3·0 per cent) if we assume:

12 An improvement in the efficiency of industrial and domestic energy use giving a reduction by 2020 of the order of 30 per cent on the oil constrained, high growth projection H4, which is equivalent to an annual reduction of 0·8 per cent over a 40-year period compared with the H4 trend. The resulting projection, denoted H5, will be described as the 'energy constrained' high growth projection. These efficiency improvements are assumed to be in addition to those arising from the 'normal' price response which is already included in H4. Whether the degree of adjustment should be considered very large (in comparison to long run productivity trends), or very small (because it involves only a small annual percentage change), depends on information which either does not exist, or is sparse, or is dispersed. This exercise shows the importance of that information. Some of the potential for adjustments towards 'energy constraint' can be assessed from savings through measures for energy conservation that have already been identified, but caution is required since many measures which are clearly cost effective will already be contained implicitly in the historical trends and in the price responses already considered.

Assumptions concerning potential energy supply

The magnitude and the necessity of the long run improvements in efficiency stated in assumption 12 depend on an assessment of the future potential for energy supply. This will be discussed in reports by the WEC supply and resource groups. However, guidelines were required for the demand projections.

13 The guidelines for the potential supply of energy are summarised in Exhibit 10. The ranges shown are intended to reflect not only the constraints for maximum potential supply indicated by the WEC resource studies but also the expectation that the potential supply for each fuel will depend on the pressure from demand. Thus potential supply will be higher for a high growth scenario than for a low growth scenario. The ranges shown are used as guidelines to influence the pattern of energy demand and fuel substitution by means of the fuel preferences in the demand model that are adjusted in an attempt to keep world demand for each fuel within the guidelines for supply. Due to the constraints noted in assumption 7, the maximum potential supply in particular fuels is not always taken up by the projections of energy demand.

14 "Fast Development" projections (FD) for the Developing Regions have been examined in combination with alternative assumptions for OECD and with the assumption that the group of centrally planned economies remains in net energy balance with the rest of the world. Two fast development projections are considered, FD4 and FD5, in which the relations between energy and economic growth are analogous to those for H4 (medium conservation) and H5 (high conservation).

15 The 'transition' scenario T4 represents an attempt to evaluate possible economic assumptions that would be consistent with:

(i) an initial period of high growth (to 2000),

(ii) medium levels of energy conservation similar to those for H4 and L4,

(iii) the guidelines assumed for potential energy supply to 2020.

Thus T4 follows the path of H4 to 2000 and thereafter economic growth assumptions are adjusted so that energy demand remains within the supply guidelines to 2020.

	Coal	Oil	Gas	Nuclear	Renewable	Total
OECD						
1972	28	28	32	2	11	101
2000	50-70	32-44	20-35	50-80	20-30	190-240
2020	70-110	26-44	10-20	100-160	35-70	280-360
Centrally Planned						
1972	33	19	12	0	13	77
2000	70-100	20-40	20-40	20-40	20-35	160-230
2020	110-150	20-40	25-45	50-100	30-60	280-360
Developing						
1972	5	67	9	0	16	97
2000	15-30	100-140	25-45	7-15	20-35	180-240
2020	20-50	70-130	30-70	30-60	30-60	220-330
World						
1972	66	114	53	2	40	275
2000	145-190	165-210	70-110	85-125	65-90	580-680
2020	220-290	135-200	70-120	200-300	105-175	820-1010

Exhibit 10. Guidelines assumed for potential energy supply
Units: Exajoules

For illustrative purposes it may be assumed that the upper and lower ends of the ranges in Exhibit 10 correspond approximately to the upper and low quartiles in a probability distribution.

3 World energy demand

3.1 Primary energy demand

Some of the results of the assumptions discussed above are illustrated by world energy demand projections shown in Exhibit 11 (alternative economic growth), Exhibit 12 (shares of world energy), Exhibit 13 (response to higher energy prices), and Exhibit 14 (improved efficiencies required to bring the constrained high projection H5 within the estimated range of potential energy supply).

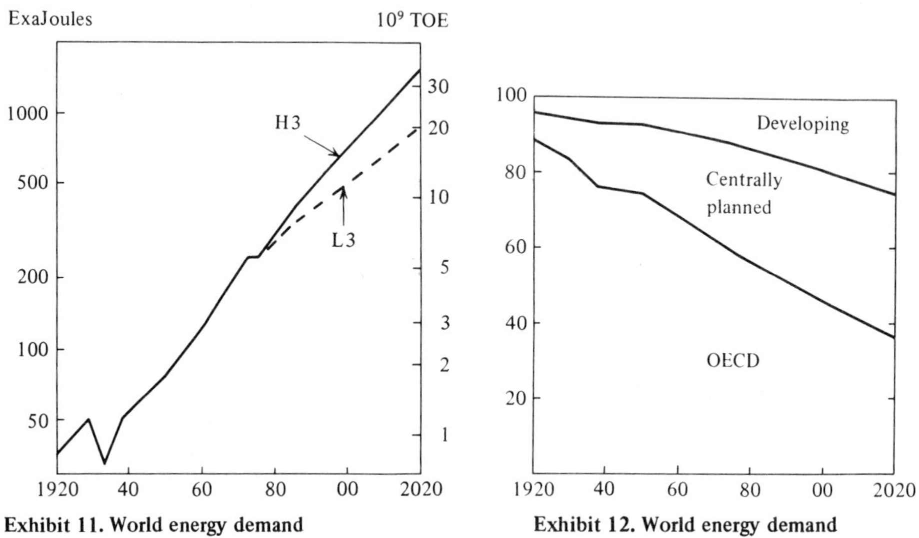

Exhibit 11. World energy demand
Unconstrained projections

Exhibit 12. World energy demand
Unconstrained high growth H3
percentage shares (excluding wood)

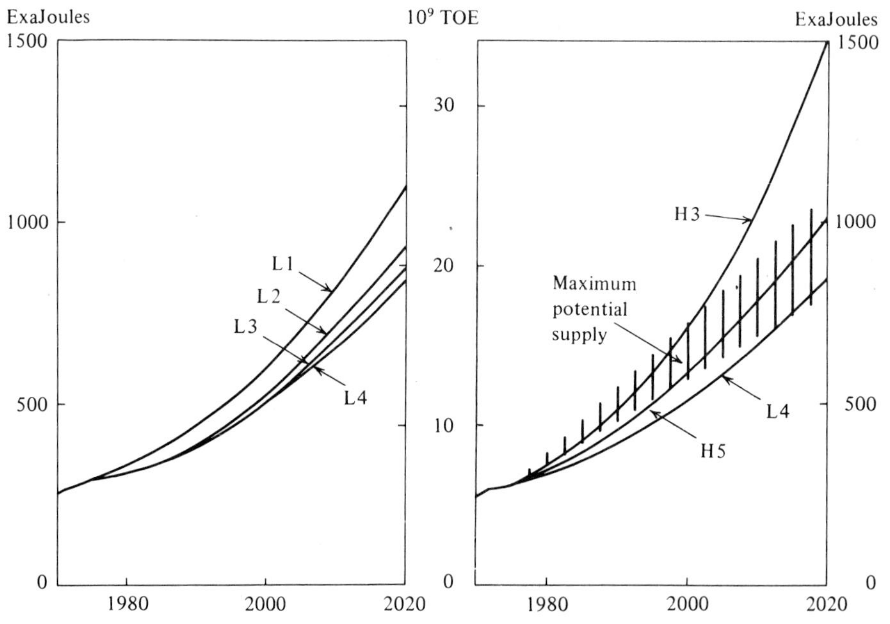

Exhibit 13. Price response of world energy demand

Exhibit 14. World energy demand and potential supply

3.2 Oil demand

At present oil accounts for over 90 per cent of the energy used in world transport. Our projections show a decline in the use of coal and an increase in electricity for transport. Results from the transport model are illustrated in Exhibit 15. The level of constraint required to keep oil demand within the range for potential supply is indicated in Exhibits 16 and 17.

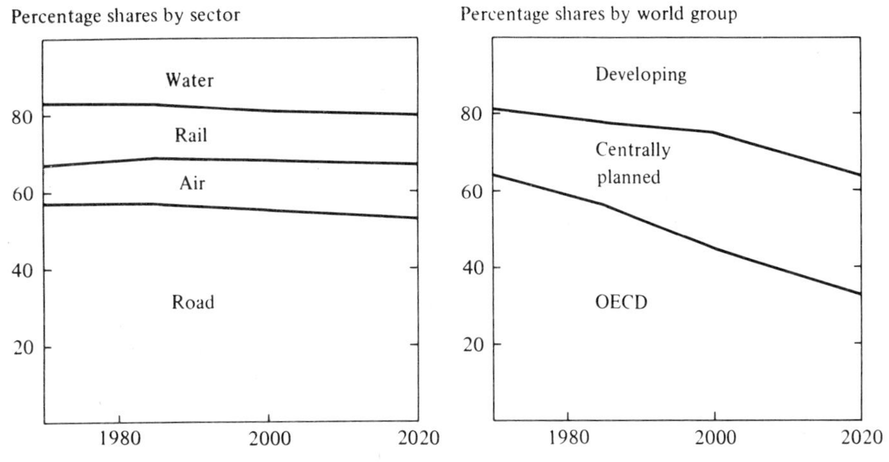

Exhibit 15. World transport energy: percentage shares

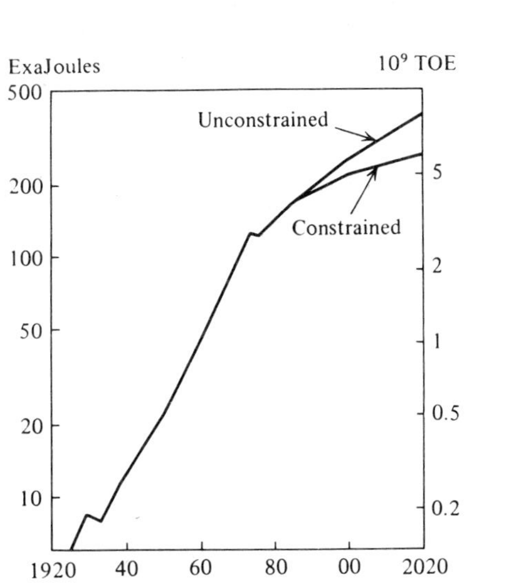

Exhibit 16. World oil demand
High growth

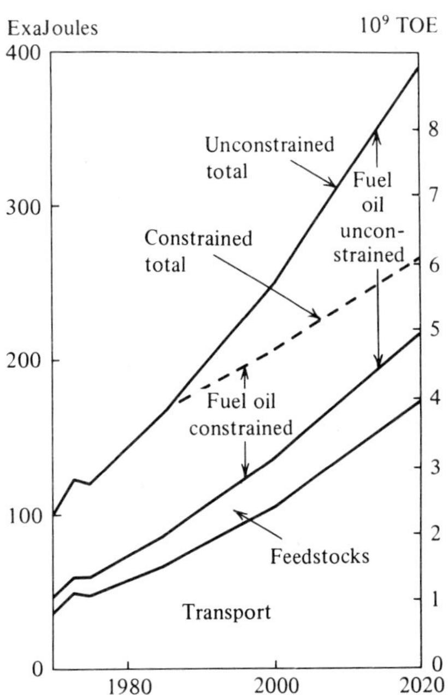

Exhibit 17. World oil demand by sectors
High growth

3.3 Fuel substitution

The principles underlying the fuel substitution model have been outlined in section 2. The model restricts maximum decline rates for each fuel in each demand sector, for example oil can decline at 5 per cent annually if it has a low preference position. Maximum growth rates are imposed on the market share of useful energy for each fuel in each demand sector, and in any 10-year period a share can increase from 0 to 7·5 per cent, or 7·5 to 10·8 and thereafter by about 4 per cent. These are not forecasts, but we find that faster rates of substitution lead to higher demand for electricity and for total primary energy, requiring both higher nuclear and higher coal supplies.

A faster increase in market shares is permitted by our model in the energy conversion sector. In 10 years a market share (for example nuclear input for electricity generation) can rise from 0 to 24 per cent, from 24 to 42, from 42 to 56. The world pattern of fuel substitution is illustrated in Exhibits 18 and 19.

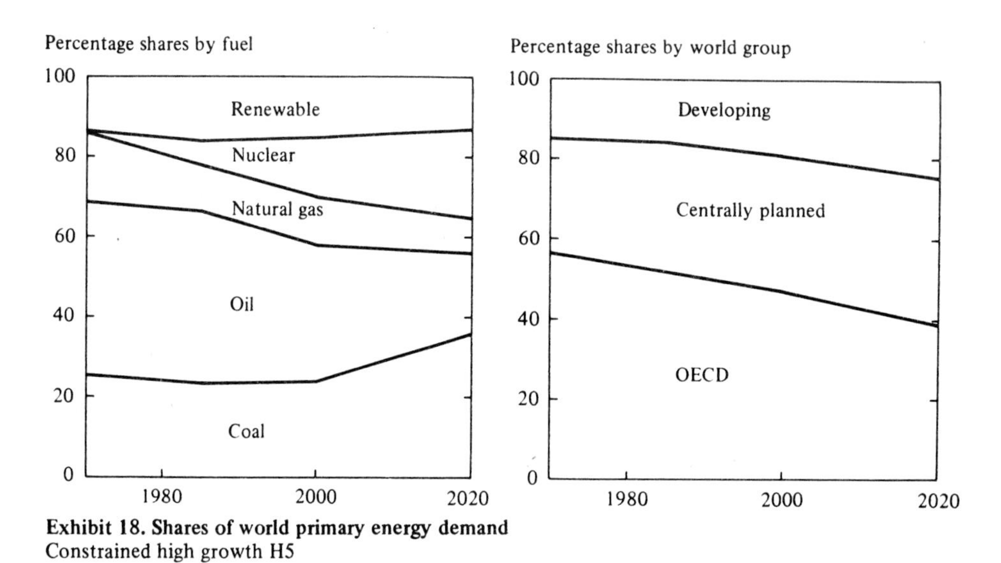

Exhibit 18. Shares of world primary energy demand
Constrained high growth H5

3.4 Energy conservation, efficiency and price response

Energy conservation normally refers to improved technical efficiencies at the point of use or conversion of energy. However, the relation between energy consumption and gross domestic product will also be affected by improved technical and economic efficiencies in areas that are quite remote from the main locations of energy consumption, for example through increased skills and the better use of labour leading to products having a higher value for the same energy input. Both kinds of improved efficiency, technical and economic, are essential to the projections of energy demand made in this report.

The modified trend projections, L1 and H1, do not contain any response to higher energy prices, but they do include the adjustment downwards of energy coefficients (energy growth relative to economic growth) that has been observed historically in later stages of industrial development. This has been assumed to take place in some of the Developing Regions in the later part of the period to 2020, and in the USSR and East Europe beginning from the mid 1970s. We would attribute such changes more to alterations in the pattern of output and structural change (for example, increasing consumer goods and services), rather than energy conservation through improved technical efficiencies.

The 'normal' price response to current and future higher prices for energy compared with other goods and services leads from the projections L1 and H1 to L2 and H2, or to L3 and H3, depending whether a medium response (15 per cent reduction over 25 years) or a high response (21 per cent) is assumed. The projections L3 and H3 would represent our low and high 'forecasts' if no thought was given to the expected scarcity of oil and subsequent difficulties in developing alternative sources of energy supply. They contain the 'normal' consumer response to higher prices leading to improved technical efficiencies and some substitution of other goods and services for energy consumption.

World Energy Demand (Exajoules)

Average annual percentage growth rates	1972-2000		2000-2020	
Economic growth	3.1-4.2		2.9-3.7	
Primary energy demand growth	2.4-3.4		2.4-3.3	
Energy demand in year	**1972**	**1985**	**2000**	**2020**
Secondary energy (EJ)				
Fossil fuel	156	199-226	263-351	339-536
Electricity	21.8	34-38	60-77	115-173
Wood, solar, etc.	26	30-33	44-57	78-116
Total final consumption	204	263-297	367-485	527-825
Primary energy (EJ)				
Coal	66	77-88	122-171	230-405
Oil	115	147-169	163-224	145-248
Natural gas	46	50-55	71-92	84-118
Nuclear	2	22-24	85-111	246-380
Hydro	14	23-23	34-34	56-56
Wood and Solar	26	30-33	44-57	78-115
Total primary energy demand	269	349-392	519-689	839-1323
Potential total energy (EJ) production	276	350-420	580-680	820-1010

Exhibit 19. World energy demand and potential supply 1972-2020

Based on scenarios L4, H5, T4 and H4.
Units: Exajoules

The potential for energy conservation has been considered in some detail for transport. It is assumed that those improved efficiencies that are technically feasible are actually achieved. Coupling these improvements with some effects from earlier saturation of demand for private motoring in developed countries and slower growth of motoring in developing countries, we move from the 'unconstrained' projection L3 to the 'oil constrained' low projection L4, giving a reduction in global energy demand of about 6 per cent. The corresponding reduction in the high growth case from H3 to H4 is about 10 per cent.

At this stage we examine the implications of fuel substitution and check whether the oil constrained projections L4 and H4 lie within the estimated ranges for energy supply. The low growth case L4 meets this criterion and therefore represents a possible scenario. Three features may be noted: firstly, the average world economic growth for the low projection is less than the historical average over the past 50 years; secondly, the 'energy conservation' reduction from the modified trend L1 to the 'oil constrained' projection L4 amounts to 27 per cent over the 48 years from 1972 to 2020; thirdly, even with low economic growth and energy savings there are pressures on energy supply and our L4 scenario requires world coal production to increase from 66 ExaJoules in 1972 to 278 ExaJoules in 2020.

For high economic growth, the 'oil constrained' projection H4 is outside the estimated range of supply near the year 2000, and lies about 25 per cent above the upper end of the supply range in 2020. In order to achieve a consistent 'energy constrained' projection H5 for high economic growth it is necessary to assume that a reduction in the growth of energy demand can and will take place in the industrial and the domestic sectors that is comparable with that in the projection for transport energy. This reduction would need to be about 25 per cent of the total demand, but over 30 per cent of demand in the industrial and domestic sectors. In the domestic sector it would involve the successful application of many of the known measures for energy conservation. In the industrial sector it could arise in part from improved technological efficiencies, but it would also require a changed pattern of industrial growth in the more developed regions.

These improvements in energy efficiencies would need to be shared by all world groups if all are to achieve the high economic growth that has been assumed. Such a large reduction in energy growth compared with conventional projections may imply that many Developing Countries would be following a different development pattern from that normally assumed, involving lower growth in energy demand whilst maintaining high economic growth. Even with this assumed reduction in energy growth there would need to be substantial expansion in all forms of energy supply such that the annual world total is increased from its present value of 276 ExaJoules to about 1000 ExaJoules by the year 2020.

4 Regional energy balances

4.1 OECD

1. North America

2. West Europe

3. Japan, Australia, New Zealand

The OECD primary energy demand projections are summarised in Exhibit 20. The results for the three regions are given in Exhibits 21, 22, 23. The values indicated for indigenous energy supplies, when taken with imports from the Developing Regions, are sufficient, within the uncertainties, to give an energy balance in the OECD Regions plus the Developing Regions in the constrained high growth case H5 to the year 2020. The average annual growth in energy demand for OECD from 1975 to 2020 is 2·0 per cent in the energy constrained high growth case H5 and 1·9 per cent for the oil constrained low growth case L4. These may be compared with growth rates of 3·8 per cent for 1950 to 1975 and 1·9 per cent for 1925 to 1950.

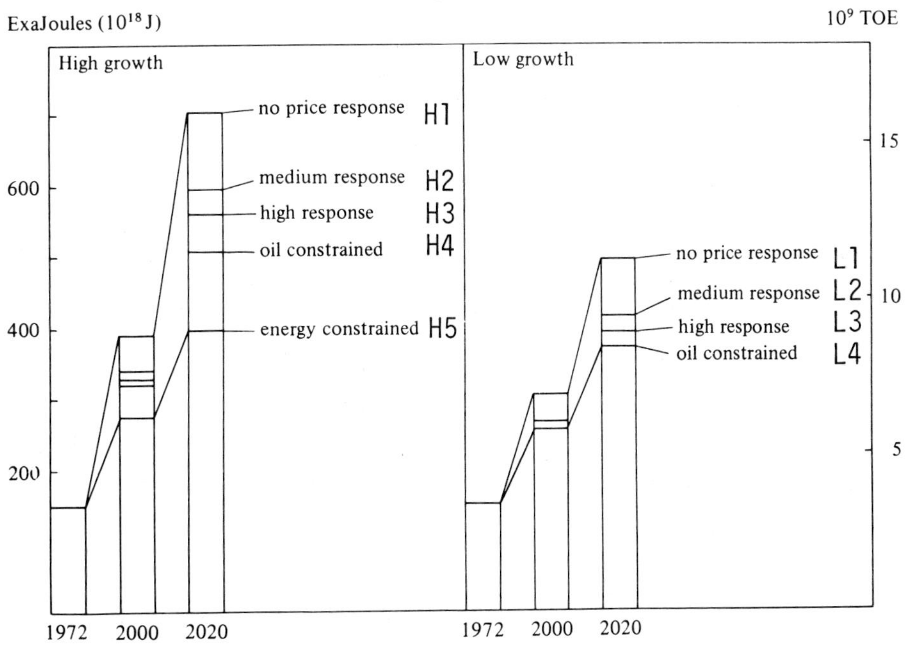

Exhibit 20. OECD primary energy demand projections

North America

The major potential for the expansion of coal production in North America makes it one of the few world regions with several options for its energy policy. In the projections it has been assumed that the region responds to world energy problems as vigorously as those regions that are less favourably endowed with energy resources. The North American results are summarised in Exhibit 21.

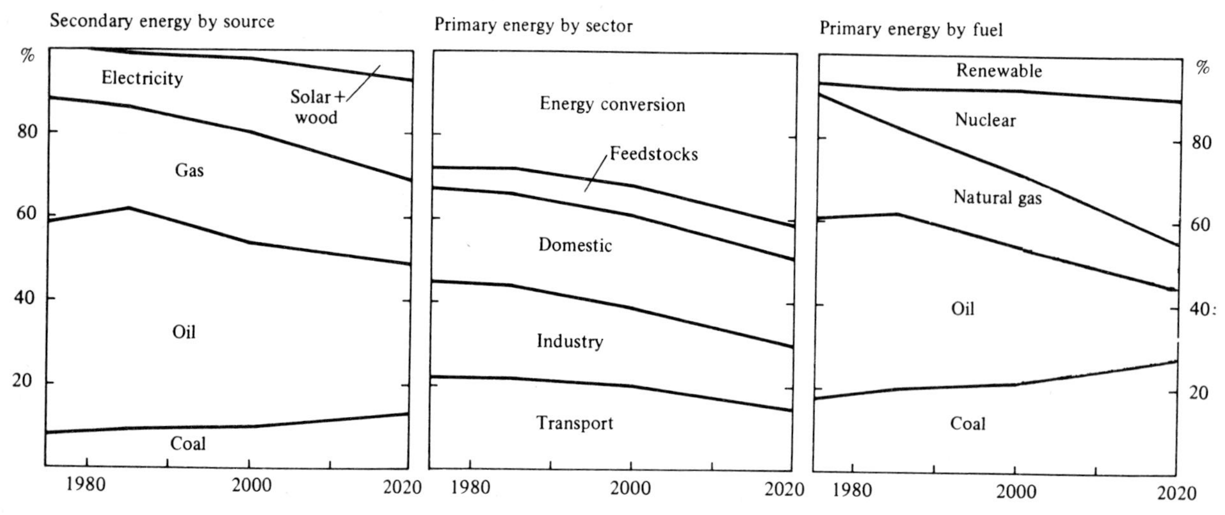

Exhibit 21. North America: energy demand
Percentage shares. Constrained high growth H5

For illustrative purposes the constrained projection H5 was used for the exhibits on percentage shares of energy demand. This projection H5 does not represent a forecast. These exhibits do not illustrate the full range of values that would be associated with the uncertainties in the projections.

North America Energy Demand (Exajoules)				
Average annual percentage growth rates	1972-2000		2000-2020	
Economic growth	2.5-3.5		2.5-3.1	
Primary energy demand growth	1.7-2.5		1.8-2.3	
Electricity growth	3.4-4.2		2.9-3.4	
Energy demand in year	1972	1985	2000	2020
Total primary energy demand (EJ)	83	97-108	134-167	193-262
Potential total energy (EJ) production	76	90-110	110-160	140-220

Based on scenarios L4, H5, T4 and H4.
The growth rates and the values for primary energy demand and electricity are indicated by ranges to emphasise the essential uncertainties in the projections. (For details on the scenarios, see the Full Report.)

West Europe

With constrained high growth H5, the projection for West Europe gives a reduction in the percentage of energy imports. They amounted to 60 per cent of total primary energy in 1972 but could be about 50 per cent in 2020 if nuclear energy is regarded as ‘indigenous’. The projection H5 is summarised in Exhibit 22.

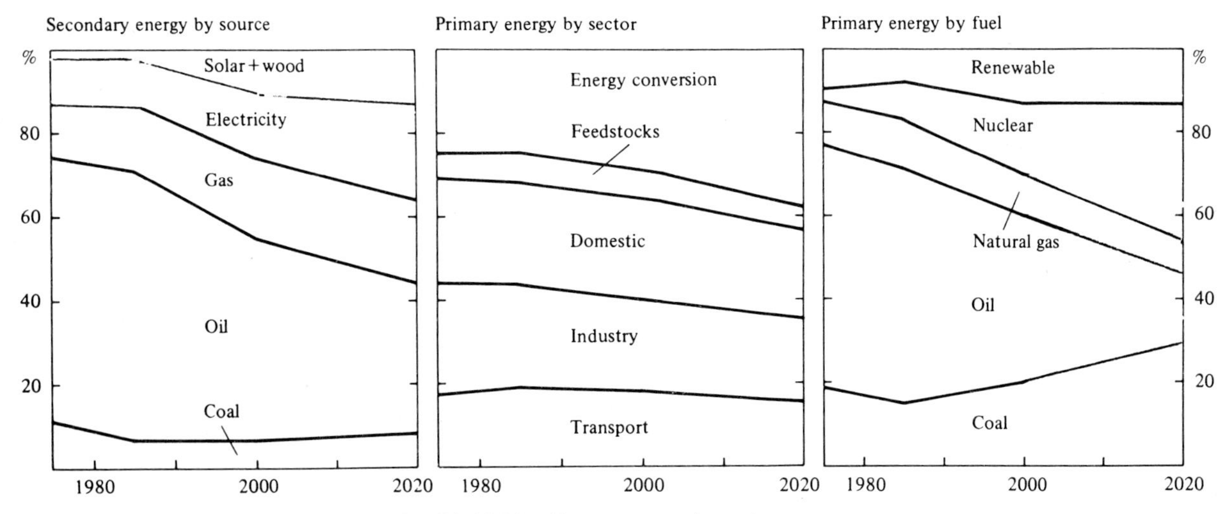

Exhibit 22. West Europe: energy demand
Percentage shares. Constrained high growth H5

For illustrative purposes the constrained projection H5 was used for the exhibits on percentage shares of energy demand. This projection H5 does not represent a forecast. These exhibits do not illustrate the full range of values that would be associated with the uncertainties in the projections.

West Europe Energy Demand (Exajoules)

Average annual percentage growth rates	1972-2000		2000-2020	
Economic growth	2.7-3.6		3.1-3.5	
Primary energy demand growth	2.1-2.8		1.6-2.2	
Electricity growth	3.7-4.5		2.7-3.2	
Energy demand in year	1972	1985	2000	2020
Total primary energy demand (EJ)	51	61-69	91-110	124-170
Potential total energy (EJ) production	20	30-40	45-70	55-110

Based on scenarios L4, H5, T4 and H4.
The growth rates and the values for primary energy demand and electricity are indicated by ranges to emphasise the essential uncertainties in the projections. (For details on the scenarios, see the Full Report.)

Japan, Australia, New Zealand

A substantial increase in Australian coal production is assumed. Including this with nuclear power, and an increase in solar and hydropower, the projection H5 leads to a reduction in imports to the region from nearly 70 per cent of total demand in 1972 to less than 40 per cent in 2020. However, Japan would remain mainly dependent on imports for fossil fuel whereas Australia would be a large exporter. The results are summarised in Exhibit 23.

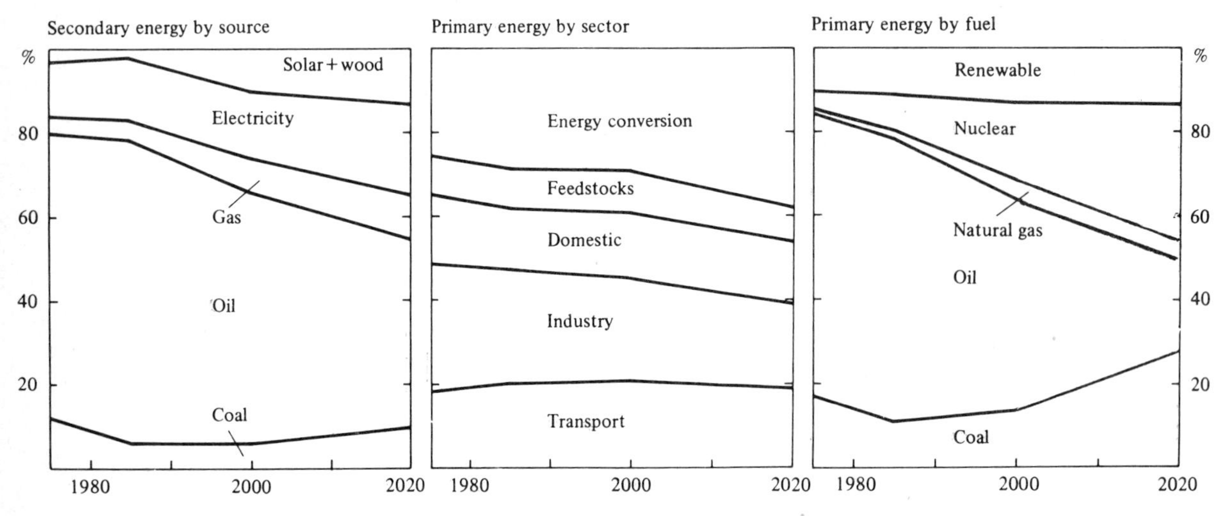

Exhibit 23. Japan, Australia, New Zealand: energy demand
Percentage shares. Constrained high growth H5

For illustrative purposes the constrained projection H5 was used for the exhibits on percentage shares of energy demand. This projection H5 does not represent a forecast. These exhibits do not illustrate the full range of values that would be associated with the uncertainties in the projections.

Japan, Australia and New Zealand Energy Demand (Exajoules)

Average annual percentage growth rates	1972-2000		2000-2020	
Economic growth	3.6-4.9		2.5-3.5	
Primary energy demand growth	2.8-3.8		2.2-2.6	
Electricity growth	4.1-5.1		2.7-3.1	
Energy demand in year	1972	1985	2000	2020
Total primary energy demand (EJ)	16	25-28	35-48	55-78
Potential total energy (EJ) production	5	7-11	20-30	35-70

Based on scenarios L4, H5, T4 and H4.
The growth rates and the values for primary energy demand and electricity are indicated by ranges to emphasise the essential uncertainties in the projections. (For details on the scenarios, see the Full Report.)

4.2 Centrally planned economies

4. USSR and East Europe

5. China and centrally planned Asia

The primary energy demand projections for the Group of Centrally Planned economies are summarised in Exhibit 24, and the results for the two regions are given in Exhibits 25 and 26. Since they form two separate trading areas it is expected that energy demand in each region should approximately match indigenous production. With the constrained high growth projection both regions are marginally in energy deficit by 2020 but both have sufficient coal reserves to increase production if this was required.

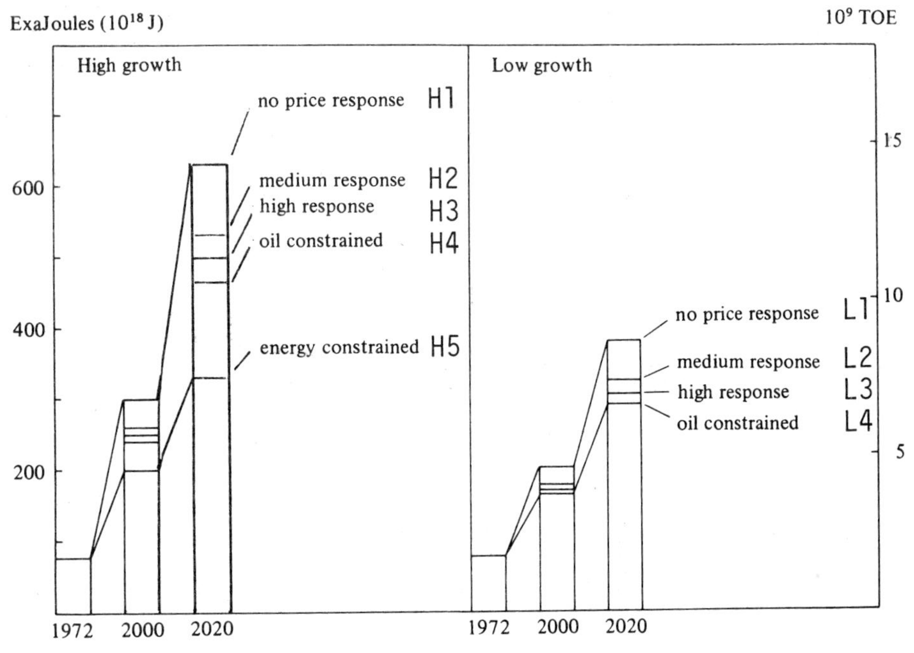

Exhibit 24. Centrally planned group: primary energy demand projections

USSR and East Europe

Historically the growth in energy demand in the USSR and East Europe has exceeded the economic growth. If the same relationship were maintained in the future the resulting energy demand would exceed our estimates for supply in this region. We have therefore assumed (as in other regions) that energy conservation and overall improvements in the efficiencies of energy use in relation to economic activity will moderate the rate of increase of future energy demand. The projected energy demand shown in Exhibit 25 for 2020 is not significantly outside the estimated range for potential supply.

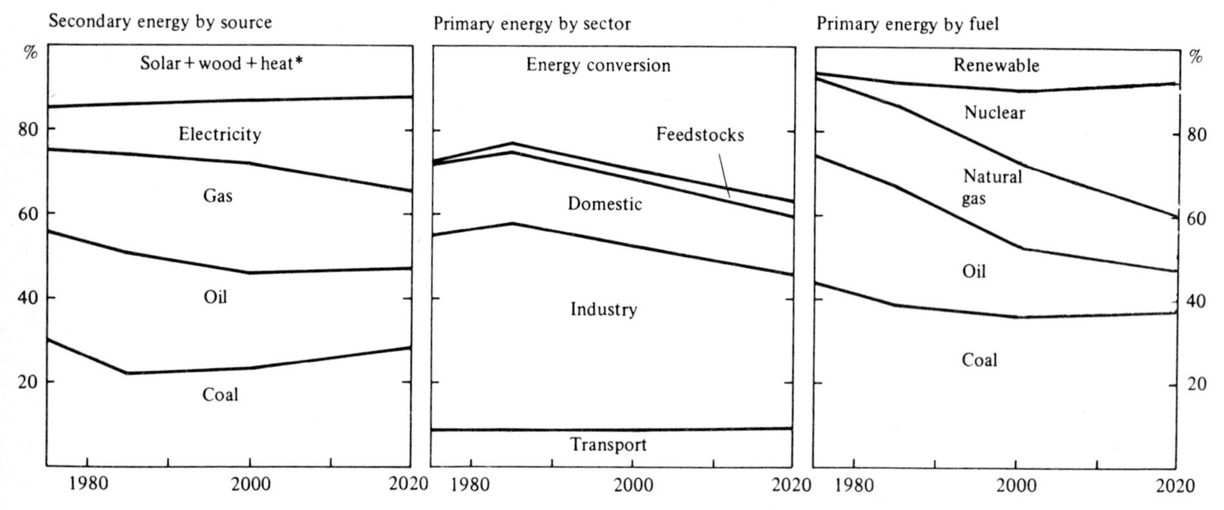

Exhibit 25. USSR and East Europe: energy demand
Percentage shares. Constrained high growth H5
* heat from cogeneration

For illustrative purposes the constrained projection H5 was used for the exhibits on percentage shares of energy demand. This projection H5 does not represent a forecast. These exhibits do not illustrate the full range of values that would be associated with the uncertainties in the projections.

USSR and East Europe Energy Demand (Exajoules)

Average annual percentage growth rates	1972-2000		2000-2020	
Economic growth	3.5-4.6		3.0-3.5	
Primary energy demand growth	2.9-4.0		2.5-3.4	
Electricity growth	3.0-3.8		3.3-3.9	
Energy demand in year	1972	1985	2000	2020
Total primary energy demand (EJ)	55	72-84	119-162	213-319
Potential total energy (EJ) production	54	60-90	110-190	150-300

Based on scenarios L4, H5, T4 and H4.
The growth rates and the values for primary energy demand and electricity are indicated by ranges to emphasise the essential uncertainties in the projections. (For details on the scenarios, see the Full Report.)

China and centrally planned Asia

Energy statistics and economic statistics are less well known for China than for other regions. However, if economic growth is assumed to continue at a high level comparable with the average for Developing Regions, the China Region would increase its energy demand more than eightfold by 2020 in the unconstrained case, and six fold in the constrained case. The projected energy demand in 2020 in the constrained case is not significantly outside the estimated range for potential supply.

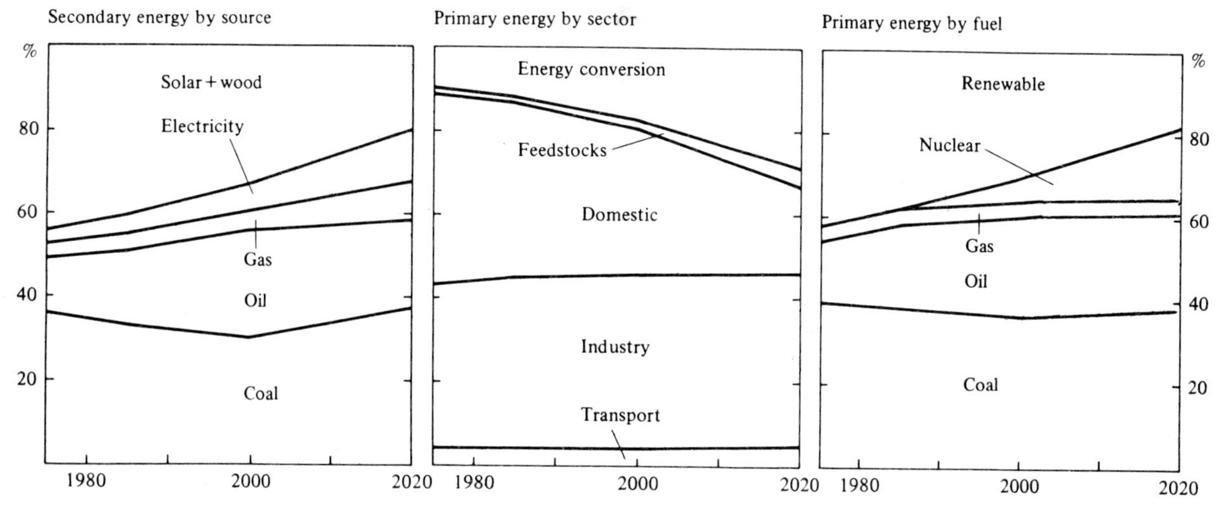

Exhibit 26. China and centrally planned Asia: energy demand
Percentage shares. Constrained high growth H5

For illustrative purposes the constrained projection H5 was used for the exhibits on percentage shares of energy demand. This projection H5 does represent a forecast. These exhibits do not illustrate the full range of values that would be associated with the uncertainties in the projections.

China and CP Asia Energy Demand (Exajoules)

Average annual percentage growth rates	1972-2000		2000-2020	
Economic growth	3.4-5.0		3.0-4.5	
Primary energy demand growth	2.5-4.0		2.5-4.1	
Electricity growth	7.5-9.2		4.8-6.3	
Energy demand in year	1972	1985	2000	2020
Total primary energy demand (EJ)	23	32-38	45-67	73-148
Potential total energy (EJ) production	23	30-50	35-70	70-120

Based on scenarios L4, H5, T4 and H4.
The growth rates and the values for primary energy demand and electricity are indicated by ranges to emphasise the essential uncertainties in the projections. (For details on the scenarios, see the Full Report.)

4.3 Developing group

OPEC Developing Group (6)
Non-OPEC Developing Group (7–11)

6. OPEC
7. Latin America
8. Middle East and North Africa
9. Africa South of the Sahara
10. East Asia
11. South Asia

The Developing Group now includes 50 per cent of the world's population but uses only 15 per cent of the energy. By the year 2020 the share of population is expected to be about 65 per cent, and the share of energy about 25 per cent. The projections for the group as a whole are summarised in Exhibit 27.

Wide variations may occur between the economic growth of different developing countries and regions, and it would be inappropriate to give estimates here. However, we require estimates for OPEC exports and must therefore examine the OPEC group. The remaining non-OPEC developing regions are studied as a single group, but their energy shares for the base year 1972 are provided and may be used to make estimates of future regional demand on a constant shares basis for the alternative high or low projections.

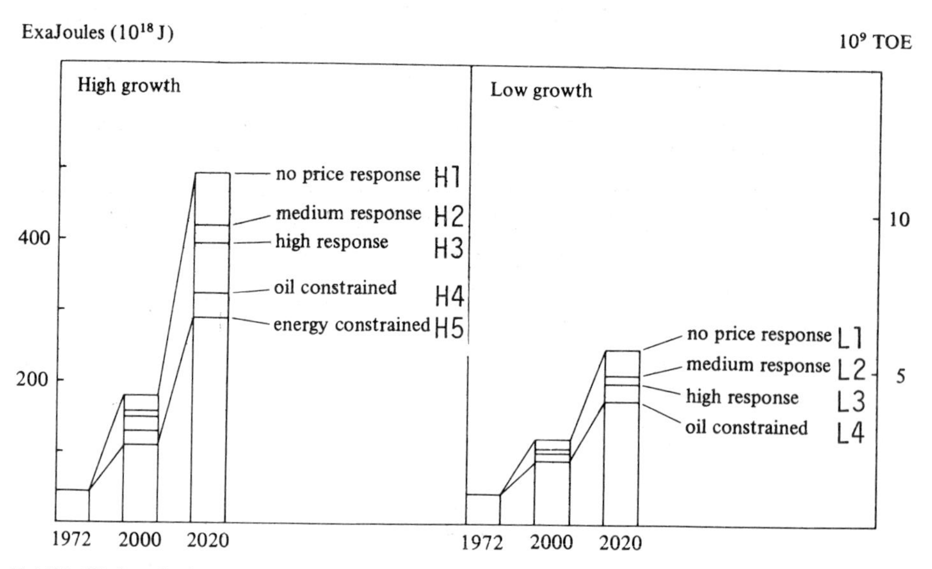

Exhibit 27. Developing group: primary energy demand projections

The energy demand for the Developing Regions in 1972 is summarised in Exhibit 28, which also gives two types of projections for the group as a whole. The first type corresponds to the range given by scenarios L4 to H4 that have been considered for all world regions, and the second to the range FD4 to FD5 with fast development in all Developing Regions. The results for the projections L4 to H4 show that the group as a whole remains a net exporter beyond the year 2000, but by 2020 the potential for exports is likely to be reduced. The fast development scenarios are unlikely to remain consistent for long after the year 2000 since the group as a whole would go into energy deficit by 2020. However, these aggregates conceal the substantial export potential of OPEC which remains until 2020 while the non-OPEC regions continue to increase their import requirements. (For details see section 5.3 of the Full Report.) The projections for the OPEC group are illustrated in Exhibit 29 and those for the non-OPEC Developing Group are given in Exhibit 30.

Energy in ExaJoules (10^{18} J) 1972	Total primary energy consumption including energy for electricity generation						Electricity output (net) ExaJoules
	Coal	Oil	Gas	Wood	Hydro	Total	
6. OPEC	0·1	3·2	1·6	1·6	0·2	6·7	0·12
7. Latin America	0·4	6·7	1·1	1·7	0·9	10·8	0·54
8. Middle East and North Africa	0·2	1·8	0	0·5	0·1	2·6	0·10
9. Africa South of Sahara	1·8	1·3	0	4·3	0·2	7·6	0·26
10. East Asia	0·5	2·6	0·1	0·9	0·1	4·2	0·23
11. South Asia	2·2	1·4	0·2	5·6	0·4	9·8	0·20
Total energy demand 1972	5·2	17 0	3·0	14·6	1·9	41·7	1·5

Developing Group Energy Demand (Exajoules)

Average annual percentage growth rates	1972-2000		2000-2020	
Economic growth	4.0-5.5		3.6-4.9	
Primary energy demand growth	3.0-4.3		3.2-4.7	
Energy demand in year	1972	1985	2000	2020
Total primary energy demand (EJ)	42	62-71	97-138	183-346
Potential total energy (EJ) production	97	125-150	180-240	220-330

Based on scenarios L4, H5, T4 and H4.

Developing Group Energy Demand (Exajoules)

Average annual percentage growth rates	1972-2000		2000-2020	
Economic growth	6.5		5.9	
Primary energy demand growth	4.5-5.3		5.3-5.9	
Energy demand in year	1972	1985	2000	2020
Total primary energy demand (EJ)	42	70-77	143-175	400-552
Potential total energy (EJ) production	97	125-150	180-240	220-330

Based on fast development scenarios FD4 and FD5.

Exhibit 28. Energy demand in the developing regions.

OPEC (region 6)

The projections for OPEC are summarised in Exhibit 29. If energy demand is 'constrained' in OPEC as has been assumed for all regions in case H5, there will be substantial net exports in 2020 around 100 EJ from the group as a whole, probably including coal as well as oil and gas. However, if OPEC was to follow the 'unconstrained' high growth path H3 instead of H5, its energy demand could be more than twice as high in 2020 and exports could be less than 30 EJ.

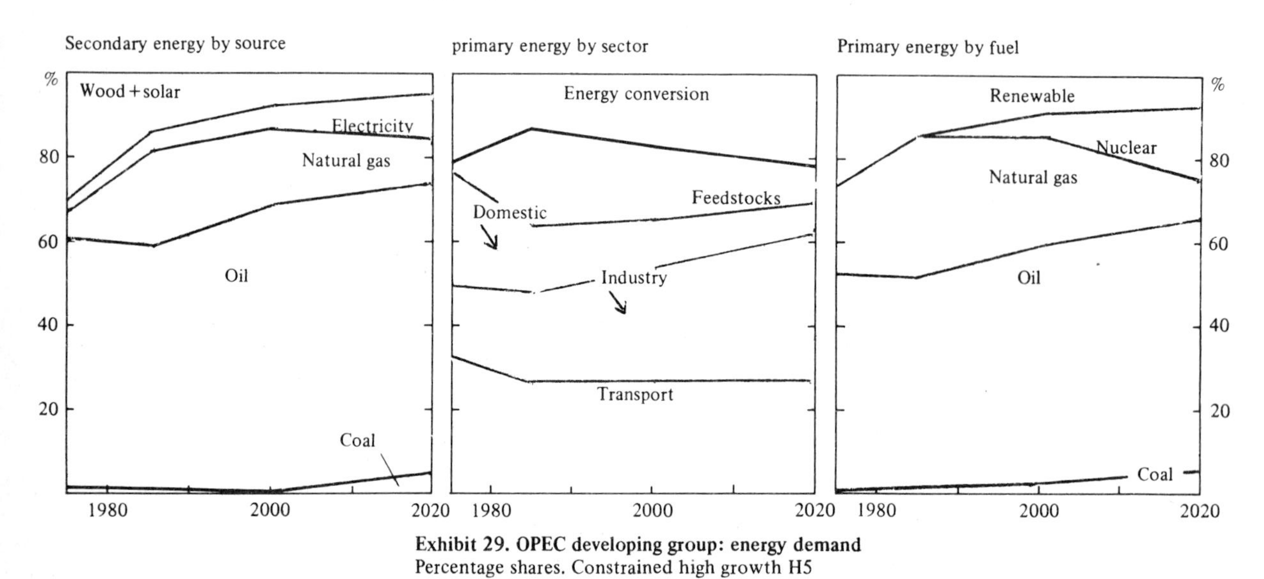

Exhibit 29. OPEC developing group: energy demand
Percentage shares. Constrained high growth H5

For illustrative purposes the constrained projection H5 was used for the exhibits on percentage shares of energy demand. This projection H5 does not represent a forecast. These exhibits do not illustrate the full range of values that would be associated with the uncertainties in the projections.

OPEC Developing Energy Demand (Exajoules)

Average annual percentage growth rates	1972-2000		2000-2020	
Economic growth	4.8-6.8		3.8-5.5	
Primary energy demand growth	3.6-5.7		2.8-4.8	
Electricity growth	7.2-9.7		5.5-7.5	
Energy demand in year	1972	1985	2000	2020
Total primary energy demand (EJ)	7	12-16	19-33	32-79
Potential total energy (EJ) production	70	80-120	100-160	90-180

Based on scenarios L4, H5, T4 and H4.
The growth rates and the values for primary energy demand and electricity are indicated by ranges to emphasise the essential uncertainties in the projections. (For details on the scenarios, see the Full Report.)

Non-OPEC developing group (regions 7–11)

If the non-OPEC Developing Group follows the high economic growth, energy constrained projection H5 their energy import requirements would remain at about 20 per cent of their total energy demand. This assumes substantial achievements in energy conservation during a period of rapid development. It also assumes a major expansion in indigenous energy production, particularly coal and nuclear power. The results are summarised in Exhibit 30.

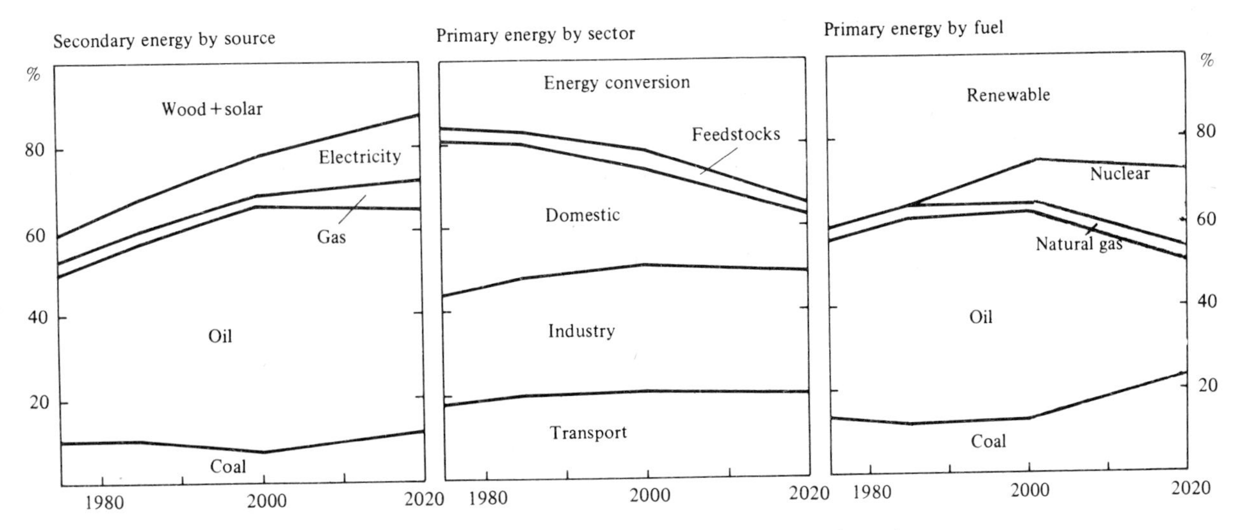

Exhibit 30. Non-OPEC developing group: energy demand
Percentage shares. Constrained high growth H5

For illustrative purposes the constrained projection H5 was used for the exhibits on percentage shares of energy demand. This projection H5 does not represent a forecast. These exhibits do not illustrate the full range of values that would be associated with the uncertainties in the projections.

Non-OPEC Developing Energy Demand (Exajoules)

Average annual percentage growth rate	1972-2000		2000-2020	
Economic growth	3.8-5.0		3.5-4.6	
Primary energy demand growth	2.8-4.0		3.5-4.8	
Electricity growth	5.5-6.8		4.8-6.0	
Energy demand in year	1972	1985	2000	2020
Total primary energy demand (EJ)	35	50-55	78-105	151-266
Potential total energy (EJ) production	27	35-50	60-100	90-180

Based on scenarios L4, H5, T4 and H4.
The growth rates and the values for primary energy demand and electricity are indicated by ranges to emphasise the essential uncertainties in the projections. (For details on the scenarios, see the Full Report.)

National Committees of the World Energy Conference

Algeria
Arab Republic of Egypt
Argentina
Australia
Austria
Bangladesh
Belgium
Brazil
Bulgaria
Canada
Chile
Colombia
Costa Rica
Cuba
Czechoslovakia
Denmark
Ecuador
Ethiopia
Finland
France
Germany (Democratic Republic of)
Germany (Federal Republic of)
Ghana
Great Britain
Greece
Hungary
Iceland
India
Indonesia
Iran
Ireland
Israel
Italy
Ivory Coast
Japan
Jordan
Korea (Republic of)
Liberia
Luxembourg
Malaysia
Mexico
Morocco
Nepal
Netherlands
New Zealand
Nigeria
Norway
Paraguay
Pakistan
Peru
Philippines
Poland
Portugal
Romania
Senegal (Republic of)
Sierra Leone
South Africa (Republic of)
Spain
Sri Lanka
Sudan (Democratic Republic of the)
Sweden
Switzerland
The Territories of Taiwan, Kinmen, Matsu, and Penghu of the Republic of China
Tanzania
Thailand
Trinidad and Tobago
Tunisia
Turkey
Uganda
Union of Soviet Socialist Republics
United States of America
Uruguay
Venezuela
Vietnam
Yugoslavia
Zambia